in Agricultural Development

Publications of the
Political Science Association
in Uppsala
67

Michael Ståhl

Ethiopia: Political Contradictions in Agricultural Development

Distributed by
Rabén & Sjögren
Stockholm

The cover photograph shows Ethiopian peasants ploughing (photographer: Lennart Isaksson).

ISBN 91–29–43743–1

Printed in Sweden by
LiberTryck, Stockholm 1974
Phototypesetting by
TEXTGRUPPEN I UPPSALA AB

Preface

This study, which is my doctoral dissertation, has been made possible by the aid and encouragement from a number of institutions and persons. A scholarship from Uppsala University and travel grants from the Scandinavian Institute of African Affairs and "Statens råd för samhällsforskning" (the Swedish Council for Social Research) have made it financially possible to carry out the study. I express my gratitude to these institutions. Furthermore I want to thank those persons who have assisted me during the work. First of all, I want to thank those Ethiopians and *ferenjis* who, during my visits to Ethiopia in 1971, 1972 and 1973, have supplied me with information, helped me to get access to source material and discussed the problems with me. I also want to thank the members of the "Internationella specialseminariet" at the Department of Political Science at Uppsala University, and especially its tutor, docent Lars Rudebeck, for the constructive criticism they have given to the ideas, papers and drafts which are integrated in this study.

Uppsala, July 31st 1974 — Michael Ståhl

Contents

Chapter 1 Introduction

This study deals with the societies of the Ethiopian highlands. Here the overwhelming majority of the population is involved in agriculture, agricultural production forming the basis of economic and political life. The emphasis of the study is on the description and analysis of concrete structural relationships defining the position of men in society. In order to place the analysis in the proper framework some general theoretical remarks on agricultural societies will be made as an introduction.

Historical and archeological research has been able to trace a positive correlation between, on one hand, the transition from hunting and foodgathering to agriculture and animal husbandry and on the other hand the emergence of politically and economically differentiated societies.[1] The first mechanization of agriculture (the animal-drawn plough) together with improved cultivated crops brought about a considerable increase in food production. The occurrence of surpluses meant that every man no longer needed to spend all his time earning a subsistence. A social order based on a division of labour and power had been rendered possible.

In agriculturally-based societies the importance of peasant cultivators is crucial. Wolf defines a peasantry thus:

> "peasants . . . are rural cultivators whose surpluses are transferred to a dominant group of rulers that uses the surpluses both to underwrite its own standard of living and to distribute the remainder to groups in society that do not farm but must be fed for their specific goods and services in turn."[2]

From the many discussions of peasantries in anthropological literature one characteristic will be stressed here: the structural subordination of the peasantry to external forces.[3] An obvious but important consequence of this is, in terms of research approaches, that a peasantry should not be studied in isolation but always in relation to the political system within which it is integrated.

The Ethiopian highlands have, during the centuries, witnessed the rise and fall of many kingdoms and chieftaincies, which have been agriculturally-based, differentiated societies. In this study some of these societies will be scrutinized; chapter 2 deals with the medieval Abyssinian kingdom and chapter 3 with the emergence of the Ethiopian empire in the nineteenth century. This historical exposé is intended to serve as a background for the ensuing chapters which deal with the post world war II land and agricultural policies of the Imperial Ethiopian Government and their effects upon the living conditions of the peasantry in Ethiopia's southern provinces, as mirrored in a few case studies.

The analysis focuses on changes in the social structure and their political background. The term social structure will be used in a broad sense. Its fundamental aspects are considered to be the relations of production. Consequently, in the Ethiopian context the concrete objects of investigation will be the pattern of landownership (the property relations) and the different forms of paying tribute (the mechanisms of surplus appropriation).

For the classification of the historical societies in the Ethiopian region, I shall draw upon the terminology elaborated by Samir Amin in relation to the modes of production. For Amin, the relations of production are the defining elements in his typology of modes of production which include the primitive community mode of production, the tributary mode of production, the feudal mode of production, the slave-based mode of production, the small scale trading mode of production and the capitalist mode of production.[4] For my analysis *the tributary mode of production, the feudal mode of production* and *the capitalist mode of production* are relevant.

According to Amin, a tribitary mode of production refers to a structural relationship where peasant communities are in the possession of the land they till, but their production is controlled by outside rulers who appropriate portions of peasant surpluses by exacting a tribute. The tributary mode of production becomes feudal when the land itself is appropriated by the rulers and the peasants are reduced to tenants, each peasant family becoming directly subordinated to a landowner. The peasant communities can no longer protect their individual member families. In the feudal mode of production, surplus appropriation still takes place under tribute forms. The capitalist mode of production is characterized by proletarization (the producers are separated from their means of production) and by commercialization (the produce as well as the labour become commodities). According to Amin there are always several modes of production in a concrete society ('social formation' in Amin's terminology), one mode of production being dominant.[5] An attempt will be made to use these analytical categories as tools in this study.

It has been stressed (by marxist authors) that the term 'underdevelopment' should be reserved for the phenomenon where the masses have been impoverished due to the imposition of capitalist relations of produciton into the third world countries. These relations have been imposed by the centre in the international capitalist system in which the countries of the third world constitute the dependent periphery. International capitalist interests, sometimes in collaboration with local powerholders, graft capitalist relations of production onto the existing pre-capitalist forms of exploitation of the people.

Peasant agriculture has been affected in several ways by this imposition. In some places peasant agriculture has been neglected due to labour migrations to, for example, the mining sector (southern Africa). In other places peasants have been alienated from their land due to the introduction of commercial plantations or large farms (Algeria, Kenya).

Studies of the social structure based upon this concept of underdevelopment have been carried out in the coffee-region of southwestern Ethiopia and in the Awash valley.[6] This study provides material on the beginning process of underdevelopment in the cereal producing highlands in the southern half of Ethiopia where private ownership of land prevails. Here local power-holders (landlors and bureaucrats) as well as peasant cultivators have responded to a 'green revolution' (fertilizer-seed-machinery package) introduced in the 1960's by the Imperial Ethiopian Government in collaboration with international agencies. The contradictions in the execution of the agricultural 'development' policies is the main theme to be analyzed in this study.

My method has been similar to that of an analytical journalist. For the analysis in chapters 2 to 4 I have relied upon available secondary source material (interpretations of chronicles, inscriptions, travellers' accounts and archeologic data). The scholars whose works I have utilized include Abir, Beckingham, Doresse, Haberland, Huntingford, Lewis, Littmann, Pankhurst, Perham, Sergew Hable Sellassie, Taddesse Tamrat and Trimingham. The only primary source which I have utilized directly is Gabra Selase's chronicle in deCoppet's French edition. I have utilized the information they give for the purposes of my own analysis. It should be added that much archeological, paleobotanic, comparative linguistic, hagiographical etc. research remains to be done in Ethiopia. Future research may therefore to a great extent complement and/or revise the present limited body of knowledge which is available on historical societies in the Ethiopian region.

For the chapters dealing with the postworld war II Ethiopian land and agricultural policies I have been able to utilize a variety of primary sources. They can be divided into three main groups, i) official policy documents, ii) research reports and survey investigations and iii) oral interviews.

The first group includes drafts and promulgated legal texts, development plans and official speeches. I have utilized these sources for a description of Government intentions and an analysis of interests underlying these intentions.

The second group of sources includes reports based on investigations of demographic, economic and social conditions in specific parts of Ethiopia. Some of the investigations have been undertaken by official bodies such as the Central Statistical Office and various ministerial departments, while others have been conducted by Ethiopian students and presented in student papers. The quantitative data presented in the reports rest on a fragile basis. No comprehensive census has ever been conducted in Ethiopia. All statistical aggregates published are estimates, often of a very crude nature.[7] In the absence of alternatives, I have quoted and used data from reports of this kind. The figures presented should, then, under no circumstances be regarded as being correct in detail. But I have reason to believe that they are indicative of basic structural relationships and can thus be utilized for a qualitative discussion.

Finally I have utilized oral interviews for two purposes; for checking data presented in written reports and for obtaining information in fields where no other sources have been at my disposal. My informants have mainly been governmental employees at different levels of the administrative hierarchy. In cases when the information has contained quantitative data (for example percentages of tenants in a certain area, rate of eviction etc.) it is based on estimates.

The analysis covers the period until December 1973. Material concerning the armed forces' intervention in Ethiopian politics, which began early in 1974, has not been included.

Notes to chapter 1

[1] Grahame Clark and Stuart Piggott, *Prehistoric Societies* (1970).

[2] Eric R. Wolf, *Peasants* (1966), pp. 3 f.

[3] For a discussion on definitions of peasantries see Sidney Mintz, A Note on the Definition of Peasantries, in *The Journal of Peasant Studies,* Vol. 1, No. 1, 1973.

[4] Samir Amin, *Le Développement Inégal. Essai sur les Formations Sociales du Capitalisme Périphérique* (1973), pp. 9—17. For a summary in English of Amin's typology of modes of production see Samir Amin, Underdevelopment and Dependence in Black Africa, in *Journal of Peace Research,* No. 2, 1972, pp. 107 ff.

[5] Amin (1973), *op. cit.,* p. 12.

[6] For an account of the pattern of underdevelopment in the coffee region in Kafa province see Jan Olof Lundberg, *Haro — ett stycke u-landsverklighet, ett samhällsreportage om utveckling och underutveckling i Etiopien* (1974), ("Haro—a Report on Development and Underdevelopment in Ethiopia"). For an account of the pattern of underdevelopment in the Awash valley see Lars Bondestam, *Den dömda dalen — om människor och kapitalism i nordöstra Etiopien.* (1974). For an English summary of this book, written in 1973 see Lars Bondestam, *People and Capitalism in the Awash Valley—Ethiopia* (1973).

[7] For an evaluation of Ethiopian statistics see Lars Bondestam, *Some Notes on African Statistics—Collection, Reliability and Interpretation* (1973).

Chapter 2 Historical Background

2.1. Prehistory and Early History

The Ethiopian region, a geographical entity, is located in the Horn of Africa. It is composed of series of mountain ranges and highland plateaus* divided by river valleys and hot arid lowlands. The rainfed highlands have a history of human settlement which traces its origin far back into the unrecorded past.

Peoples speaking different Kushitic languages inhabited the Ethiopian region at least as early as the second millenium B.C. Scholars have subdivided Ethiopian peoples speaking Kushitic languages in various fashions, the Beja, Agaw, Sidama, Afar, Somali and Oromo being some main groupings.[1] In early times the Beja inhabited the lowlands in the north, the Agaw inhabited the Tigrean plateau and the central massif, the Sidama inhabited the Shoan plateau and the southwestern plateau and the Afar, Somali and Oromo inhabited the Awash basin, the southeastern highlands and their adjacent plains.[2]

During the first millenium B.C. a movement of people into the Ethiopian region took place. Semitic speaking tribes from southern Arabia crossed the Red Sea and pushed into the highlands where they mingled with the Agaw.[3] The Semitic incursions brought political and commercial dynamism into the region. It has been stated that the emergence of differentiated societies in the Ethiopian region was conditioned by contacts across the Red Sea[4]. But before turning to the development of the kingdom of Aksum and its successor the Abyssinian kingdom, some economic preconditions will be discussed.

The economy of the Beja, Somali, Afar and Oromo was pastoral, cattle, camels and goats being the major assets of the clans.[5]

The Agaw and Sidama highlanders subsisted on agriculture and animal husbandry. The origin of agriculture in the Ethiopian highlands has been a controversial issue in the scholarly debate. One theory claims that Ethiopia constituted an independent centre of origin of cultivated plants.[6] It has also been argued that the Agaw were introduced to early forms of agriculture through impulses from the Nile valley, while the immigrants from Arabia

* The highlands include the Tigrean plateau, the central massif, the Shoan plateau, the southwestern plateau, the Harar plateau and the Bale-Arussi massif. See map 1.

brought with them the plough and modified agricultural technique suitable to cereal cultivation.[7] Another hypothesis is that the Agaw practised cereal plough cultivation prior to the influx of the Semites.[8] Lack of data makes it impossible to verify any of these hypotheses. Here it is sufficient to conclude that the Ethiopian highlands have a long tradition of agricultural (mainly cereal) production. The animal-plough technique, which in any case was practised in the first millenium B.C., indicates that the cultivators were capable of producing a significant surplus of food stuffs (teff,* barley, wheat), a precondition for the emergence of differentiated societies with central political authority.

The nature of the relations which were established between the Agaw population and the immigrants is not known. The possibility that the Agaw were already organized in small kingdoms before the arrival of the immigrants cannot be overlooked.[9] It has been assumed, however, that the latter settled among the original population and took the political initiative which resulted in a number of independent political entities in the plateau region.[10] The relations of these small kingdoms with Arabia declined in favour of a flourishing Mediterranean trade, first with Hellenic Egypt and later on with the Roman and Byzantine empires.[11] In the Red Sea ports below the Tigrean plateau commodities such as ivory, gold, rhinoceros horn, tortoise shells, and slaves were exchanged for iron weapons, copper, cloth etc.[12] This Mediterranean oriented trade stimulated political centralization and in the first centuries A.D. one kingdom centred in Aksum emerged as the supreme political power of the Tigrean plateau. The kingdom consisted of a core around the town of Aksum and a number of more remotely located areas whose kings or chiefs paid tribute to the Aksumite king.[13]

During its position as a great power (fourth century to seventh century) Aksum influenced a large area stretching north to Nubia, west to the Atbara river, south into the central massif and east across the Red Sea into the coastal regions of Yemen.[14] Aksum's wealth—even today reflected in impressive obelisks, temple ruins and sepulchral monuments—seems to have been founded on military control over strategic trade routes, i.e. control over Red Sea ports and caravan routes from the Red Sea to Aksum and further into the mountainous hinterland.[15] Stone inscriptions relating the warlike expeditions which Aksumite kings embarked on, indicate that the expeditions were made to secure and enlarge control over trade routes.[16]

Aksum's foremost trade partner was the Byzantine empire. Aksumite

* Teff is a cereal plant endemic to the Ethiopian region.

military power, which included a navy, secured Byzantine maritime commerce on the Red Sea and accessibility to valuable raw products.[17] Aksum received important processed goods and considerable cultural influences. One lasting cultural influence was the introduction of Christianity. In the fourth century Aksum's ruler Ezana had himself as well as his court baptized and the conversion "may not have been totally free of diplomatic and political considerations".[18] The Christian religion and the Semitic language (Geez) distinguished Aksum's cultural profile. Christianity did not reach the general population until the end of the fifth century when evangelization–encouraged by the kings–commenced on a large scale.[19]

The productive forces developed, metal tools were used, irrigation installations made and agricultural terracing practised.[20] Surpluses from agriculture together with trade revenues enabled the rulers to endow Aksum with a monumental architecture.[21]

This for Aksum favourable international situation came to an end in the seventh century, when Muslim expansion terminated Aksumite presence in the Red Sea and began spreading inland.[22] In the eighth century the pastoral Beja tribes moved on to the Tigrean plateau and cut off Aksum's access to the coast.[23] Consequently, trade revenues decreased and the power of the Aksumite kings declined.

Few details are known about the situation in the highlands from the eighth to the thirteenth century. The scanty data available indicate continuous fighting. The Mediterranean orientation was lost, and Aksum faced a new situation. With sword in hand and the gospel on their lips the late Aksumite rulers turned south to the harsh, central massif.

> "One of the most characteristic aspects of the process of expansion was the establishment of military colonies in the newly conquered territories . . . the Agäw were the native inhabitants of the Ethiopian plateau north of the river Jäma. The immigrants from south Arabia settled in the northern part of this region and lived among the Agäw population, who gradually adopted the language of the settlers. These linguistically semitized natives later constituted the dominant section of the peoples of Aksum. As the kingdom acquired additional territories, large groups of these people were apparently recruited and settled in distant frontier stations. Here they intermarried with the local people and formed yet another semitized zone in the interior of the Agäw country, which was gradually passing into the sphere of Aksumite political influence. It is apparent that all the Semitic linguistic groups south of the Tigre' region had a similar origin."[24]

2.2. The Abyssinian Kingdom

During the first half of the twelfth century a Christian Agaw dynasty called Zagwe gained control and established itself as ruler over a long and narrow corridor stretching from Aksum's old centre through the central massif down to the Shoan plateau and Awash river. The power of the Zagwe kings is reflected in the impressive rock hewn churches which they built in Lasta, the cradle of the dynasty.[25]

In the late thirteenth century the Zagwe dynasty was overthrown by an alliance composed of their political enemies in Tigre to the north and Shoa to the south. The alliance was headed by the Shoan chief Yekuno Amlak. He claimed to be a descendant of the old Aksumite kings and founded a new dynasty called the Solomonic dynasty.[26]

The territory under the rule of the Solomonic dynasty will here be called the Abyssinian kingdom*. The central regions of the emergent kingdom were Tigre in the north around the old city of Aksum, Wag and Lasta in the central massif and parts of the Shoan plateau further south.[27] In all of these regions the Orthodox Church became firmly established.[28] In Tigre the Semitic Geez language had developed into Tigrinya, while in Shoa the Amharic language (a Semitic language not necessarily directly related to Geez) was spoken. Amharic was the language of the Abyssinian kings and by the time it became widely diffused in the central massif.

From its original area the Abyssinian kingdom expanded in two main directions. One front was in the west where the battle was against independent Agaw political units.[29] The long term result was that new areas were made subject to the authority of the Abyssinian kings. A gradual christianization and semitization took place, while the Amharic and Tigrinya languages replaced the local ones.

The other front was in the southeast where the conflict was with a string of Muslim sultanates which had been established in the region with the spread of Islam from the coast to the interior.[30] The sultanates were important merchant societies which carried on trade with the rich Sidama areas where societies with monarchial institutions had developed. Throughout the fourteenth, fifteenth and sixteenth centuries wars were waged between the Abyssinian kingdom and these sultanates. The actual issue of these conflicts was control over trade and trade routes.[31] Most of the time the Abyssinian armies had the upper hand in the military conflicts and thus the Muslim sultanates were forced to pay tribute to the Abyssinian king, but

* I use the term Ethiopia to denote the geographical unit and Abyssinia to denote the political unit; one of the historical kingdoms which has existed within the Ethiopian region.

Map 1. THE ETHIOPIAN HIGHLAND REGIONS

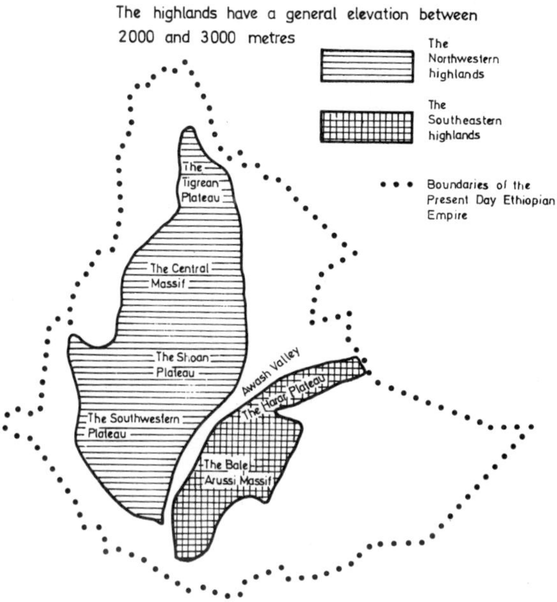

Source: Mesfin Wolde Mariam, *An Atlas of Ethiopia.*

Map 2.

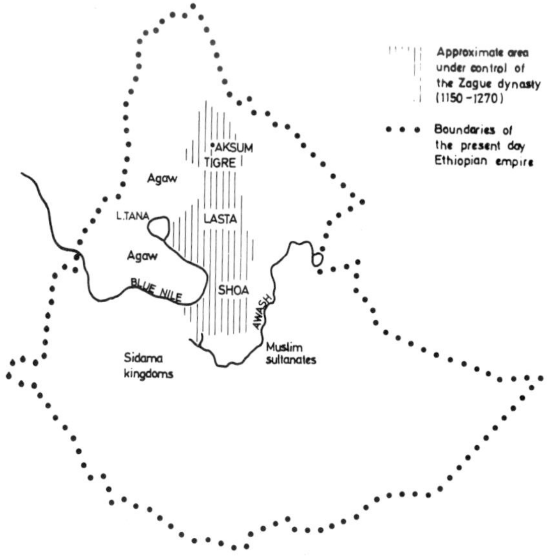

Source: Taddesse Tamrat, *Church and State in Ethiopia 1270–1527.*

MAP 3. THE ABYSSINIAN KINGDOM IN THE 14TH CENTURY

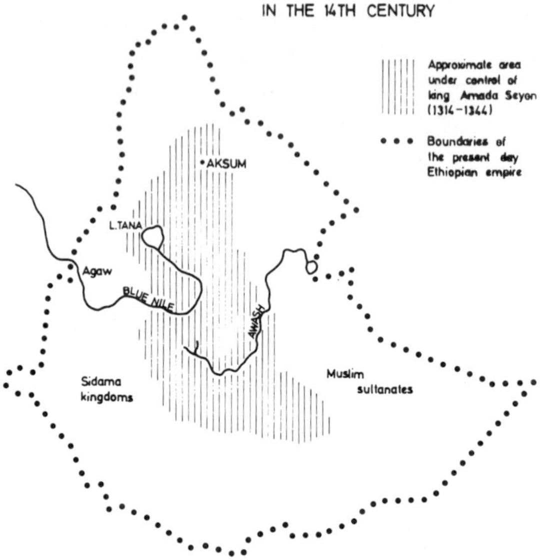

Source: Taddesse Tamrat, *Church and State in Ethiopia 1270–1527.*

they kept their political and cultural identity and were not incorporated into the kingdom.[32] During the most expansive phases of the Abyssinian kingdom the Sidama kingdoms south and west of Awash river were also reduced to tributary status, but they were not effectively incorporated with Abyssinia.[33]

The Abyssinian kingdom held the position of a great power in the Horn of Africa until the first decades of the sixteenth century. It maintained diplomatic relations with the Muslim Near East and even with Mediterranean Europe, where it was vaguely referred to as "the kingdom of Prester John". While the international political relations of the kingdom will not be dealt with here, some main characteristics of its internal social structure will be analyzed.

2.3. An Outline of the Social Structure in the Abyssinian Kingdom

Narrations by Portuguese travellers visiting the Abyssinian kingdom tell of a fertile agricultural country abundant with crops and livestock.[34]

In the kingdom land was held by peasant communities.[35] The communities were organized on a kinship basis, members of family groups claiming the right to use agricultural land within the area which belonged to the community. Land ownership was vested in a mythical founding father of the kinship. According to tradition the founding father (*akni abat*) was the man first to settle in the area and take up agriculture. The following generations continued in his footsteps and the land held by them, nominally in the name of the founding father, was called *rist* land, i.e. hereditary land.[36]

The founding fathers received ownership rights either by the mere fact of occupying and cultivating an area or through a grant from the Abyssinian king.[37] In both cases the establishment of *rist* ownership seems to originate from the conquest of Agaw areas by Tigrinya and Amharic speaking warriors.[38]

The gradual christianization and semitization of the highland regions indicate that an assimilation of the conquerors and the conquered took place. It must however be assumed that the peasant society was differentiated internally. The Abyssinian monk Bahrey gives a clue to this differentiation when writing about the different 'classes' of people in the kingdom.

Bahrey writes on one hand about "elders" who are lords and hereditary landowners and on the other hand about "labourers in agriculture who live in the fields".[39] How this internal differentiation was established is not known in detail.

Ruling over the peasants was a hierarchy of authorities, the apex of which was the Abyssinian king. For the administration of the country the king appointed regional governors. The remuneration of governors was organized so that the king waived part of his taxing power over the peasantry in favour of the governors.[40] This system was institutionalized in the form of *gult*, i.e. lands granted by royal charter to the governors.[41] *Gult* referred not to the actual ownership of land but to rights to exact tributes from the peasantry living in the area which had been chartered to the governor. The actual ownership remained in the hands of the peasant communities.[42]

The nature of the obligations of the peasantry to the *gult* holders are not known in detail, but it is selfevident that the obligations included agricultural products and labour. Among the items presented as tribute were home-made cloth, grain, honey, salt, camels, sheep and cattle. The amount of the tributes may have varied from 1/10 to 1/3 of the peasants production.[43]

In exchange for these privileges the governors were expected to maintain law and order in their areas of administrative responsibility (which were larger than their *gult*-area), collect taxes for the king and raise troops for the king's military expeditions. Two basic administrative levels existed.

There was the district level, whose governor was called *shum* (i.e. "one who has been put into office"), while a number of districts formed a province whose governor held the high title of *mesfin, makonnen, nagashi* or *sahäfä lam.*[44]

Gult privileges were in principle connected with the administration of the Abyssinian king's regional government, and were as such revokable. When a *gult* holder's administrative authority terminated, he also lost his *gult*-privileges which were then passed on to his successor.[45] At the same time as the *gult* institution was a major mechanism of appropriation of peasant surpluses, it provided the king with an instrument for political centralization in the sense of making regional authorities dependent on him. A prerequisite for this kind of centralization was that the physical authority of the king could reach all corners of Abyssinia, an ambition which the kings tried to live up to by moving around with court and army from province to province. There were also a number of royal estates in different parts of the kingdom.[46]

It must be assumed, however, that there was constant tension between the royal ambition to monopolize appointments to regional offices and the concomitant *gult* privileges on one hand and the regional governors' ambition to secure office holding and *gult* privileges within their family on the other hand. During the reign of powerful kings like Amda Seyon (1314—1344), Zara Yakob (1434—1468), Baeda Maryam (1468—1478) and Lebna Dengel (1508-1529) the principle of dismissing regional governors at the will of the king could successfully be enforced,[47] while in periods of a weak central Government the regional governors must have been more independent.

Gult privileges were also granted by royal charter to churches and monasteries and such grants were considered to have a lasting character.[48] Although the royal charters establishing *gult* rights to religious institutions generally stated a pious motive for the grant, it is obvious that political considerations influenced the grants.

> "The new Abyssinian kingdom, no longer confined to the northern part of the plateau, but in the very heart of the Ethiopian massif, was in a state of continuous warfare with the Agao peoples of the interior, whose resistance more than once menaced its very existence."
>
> "The challenge of the pagan Agao was met by the embryonic Christianity of the new kingdom in a way which resulted in the fusion of political and religious interests. The conversion of the Agao became a political necessity if the State were to survive."[49]

Viewed in such a light it is easy to understand that the kings were eager to enhance the material standard of the clergy, who performed the evangelization work.

Over a period of centuries the cathedral in Aksum and other important monasteries and churches accumulated considerable wealth. The monastery of Dabra Bizan is mentioned by a Portuguese by the name of Alvares who travelled in Abyssinia in the sixteenth century. According to him the *gult* privileges of Dabra Bizan included all land within a radius of ten leagues from the monastery as well as numerous villages even further away. Every village paid fifty cows every third year to the monastery.[50]

The political importance of the Church for the monarchy was further enhanced by the fact that it was the only educational institution in the country. Literate monks were recruited from the monasteries to the court, where, as the educated elite, they filled such important and influential positions as the king's chronicler, secretaries and advisers.[51]

In summary then, it can with a high degree of certainty be stated that the *gult* system was the major mechanism of surplus appropriation in the Abyssinian kingdom. Through the *gult* grants parts of the production emanating from the peasants' *rist* lands were channelled up to such authorities as the Church, the regional governors and the king and his court, and provided them with their material basis of existence. On the district level, where *gult* holders appear to have been members of powerful families within the local area,[52] it can be assumed that the surplus appropriation occurred under peaceful and patriarchal terms and involved no violence. Under higher authorities, such as provincial governors who due to the king's appointment policies were often strangers in the area under their administrative responsibility, the *gult* system was often detrimental to the peasants' interestes. Alvares comments on the situation in the early sixteenth century:

> "There would be much fruit and much more cultivation in the country, if the great men did not ill-treat the people, for they take what they have, and the latter are not willing to provide more than they require and what is necessary for them".[53]

de Almeida, a Portuguese traveller who visited Abyssinia one century later, describes the abuses of the policy of centralized appointments of regional governorships:

> "The worst thing is that . . . the . . . governorships of all his [the king's] kingdoms and provinces seem to be sold rather than given. No one receives them except by giving for them an amount of gold which is more or less the income and profit the aspirant and applicant hopes to get from them. As there are always many applicants, those who give most for them usually receive them. They give much more than they can honestly derive from them, and so as not to be at

> a loss, they fleece the people, disposing of the lesser offices and the governorships of particular places and territories to those who promise and give most in return for them. So it is all merely an auction. Since these governors are the lords and judges and hold in their hands absolute control over the lives and property of the whole population, generally speaking, they are all plunderers rather than governors".[54]

It can be concluded then, that the brutal application of the *gult* system tended to inhibit the productive forces. Neither the governors nor the governed were interested in agricultural innovation. The former knew that their term of office was short and the latter were continuously plundered.

Finally a few comments will be made on the question of the mode of production. Without probing deep into the debate which has followed Marx's discussion of the Asiatic mode of production I wish to take up a recent contribution made by Amin. According to Amin the analytical concept of mode of production can be utilized to characterize concrete societies. Among the pre-capitalist modes, he sketches the mode of production of the primitive village community and the tributary mode. The definition of the latter reads:

> "the tributary mode of production which involves the persistent parallel existence of a village community and a social and political structure which exploits the former by exacting a tribute".[55]

The definition is elaborated thus:

> "Le mode de production tributaire est caractérisé par la séparation de la société en deux classes essentielles: la paysannerie, organisée en communautés, et la classe dirigeante, qui monopolise les fonctions d'organisation politique de la société et perçoit un tribut (non marchand) sur les communautés rurales".*[56]

The point here is that the actual ownership of the major means of production, land, is vested in peasant communities, while these are politically subject to outside authorities who demand a share of their production.

In the Abyssinian context the essential institutions of *rist* and *gult* fit into this definition, *rist* being an expression of the peasant communities' possession of the major means of production and *gult* being an expression of the military/political and ideological authorities' mechanism for the appropriation of peasant produced surplus.

* English translation: The tributary mode of production is characterized by the separation of society into two essential classes: a peasantry organized in communities and a ruling class which monopolizes the political organization of the society and exacts (non-commodity) tributes from the rural communities.

Although landownership was formally vested in the peasant communities, agricultural production was to a considerable extent outside the control of the producers.

It can be claimed that the tributary mode of production was the dominant one in the Abyssinian kingdom.

These remarks have far from exhausted the issue. One must assume the existence of peasant communities living in isolated areas beyond the influence of the royal authorities. There remains also the questions of slavery and long distance trade. These will have to be elaborated upon before a full description of the Abyssinian kingdom in terms of modes of production can be made.[57]

Little is known about the social structure of the societies belonging to the southwestern highlands. In these Sidama areas social stratification was highly developed, medieval sources reporting the existence of powerful kingdoms here.[58] Few details are known about them and their origin is unknown. It has, however, been argued that before the advent of the Semitic speaking tribes into the Tigrean plateau, a common culture including an institution based upon kingship joined the peoples throughout the Ethiopian highland regions.[59]

2.4. Political History 1500–1855

In the sixteenth century the Abyssinian kingdom's position as the supreme power in the Ethiopian region came to an end. In the international system bordering on the Indian Ocean the interests of two new great powers, the Ottoman empire and Portugal, clashed.

Repercussions of the conflict were registered in the Ethiopian region where the Muslim sultanates were strengthened by the Turks and diplomatic negotiations were carried on between the Abyssinan and Portuguese courts.[60] Supported and encouraged by the Turks, the Muslim sultanates united and embarked on a *jihad* against the Christian kingdom. The *jihad* was successful in the beginning, the Abyssinian armies were repeatedly defeated and the Christian highlands were looted. But the invadors, the bulk of which were Afar and Somali pastoralists, lacked the organizational capacity to transform the temporary occupation into permanent dominance. The Abyssinian forces, supported by the Portuguese could counterattack and the *jihad* eventually failed.[61]

In the wake of the *jihad* Abyssinia became the object of military pressure from pastoral Oromo* tribes which, during the following centuries, penetrated deep into the kingdom through continuous battles.

The sudden demise of the Abyssinian kingdom's supremacy has been dramatically summarized by Levine:

> "Amhara-Tigre society was convulsed by one disaster after another during the succeeding century. In 1527 began a series of conflicts with the Muslim kingdom of Adal led by a fearful warrior, Imam Ahmad ibn Ibrahim al Ghazi, surnamed Gran. Equipped with firearms from the Ottoman Turks and hordes of Somali soldiers, Ahmad Gran proceeded to overrun Ethiopia [the Abyssinian kingdom],causing destruction and massacre wherever he tread. Only after most of the country had been laid waste, its inhabitants forced to embrace Islam, its churches looted and burned, was Ahmad Gran at last killed. His troops were routed by an Ethiopian force that had been revived by the expedition of Portuguese match-lockmen under Christopher da Gama.
>
> The wars with Ahmad Gran left the country impoverished and disorganized. In this state it lay vulnerable to another shock: the massive tribal movements of the Galla from the south. The Galla advanced through periodic invasions of their warrior age groups. Coming first in furtive raids, then concerted warfare, finally by slow migration, these hardy nomads managed to penetrate deeply into Amhara lands. By 1563 the Galla had invaded over a third of the empire."[62]

The rapid and warlike expansion of the Oromo, which seems to have started quite suddenly in the mid sixteenth century, has not yet received a thorough explanation.[63]

The fact remains however, that the Oromo tribes spread over a vast area of the Ethiopian region. It encompassed the southeastern highlands and their adjacent plains where they gradually reduced the Muslim sultanates to insignificance and the southwestern plateau where they conquered some Sidama kingdoms which earlier had paid tribute to the Abyssinian king. The Oromos encircled other Sidama areas. They also overran the Shoan plateau and penetrated deep into the central massif. The Oromo warriors learned from the Abyssinians to fight on horseback and their military speciality was the cavalry shock.[64]

The Oromos did not act as a unified political unit. The rapid geographical expansion in different directions created economic and religious subdivisions between them. Some groups remained pastoral while

* In the literature the name Galla is generally used to denote the Oromo tribes. The word Galla is however a nickname with derogatory implications. I therefore prefer to use the word Oromo, which is used by the Oromos themselves.

others, who conquered agricultural areas, became settled and shifted their emphasis to an agricultural economy; some groups remained pagan, while others embraced Islam and still others Christianity.

The effect of the Oromo incursions on the Abyssinian territory was that the energy and resources of the kings continuously had to be directed toward checking and counteracting the Oromo offensive. In this process the kingdom was interspersed with corridors of Oromo settlement, while the king's army could not reach all regions of his territory. In the eighteenth century there crystallized a number of Abyssinian regions. The principal regions were Tigre, Beghemder, Lasta, Godjam and Shoa. The king resided in Beghemder where the capital of Gondar had been established, while the other regions were under the rule of warlords of varying magnitude. The monarchy was weakened, real power being vested in the regional warlords.[65]

The monarchy collapsed completely in the mid eighteenth century when Oromo chiefs, through military alliances and intermarriage, managed to infiltrate the court of the Solomonic dynasty and became its arbiters, thus completely alienating the Amhara and Tigre warlords in the Christian regions.[66] The period between 1769 and 1855 is called *zemene mesafint*, i.e. the era of the princes, and was characterized by endless wars between changing constellations of Abyssinian and Oromo warlords who used the kings as puppets.

The wars severely affected agricultural production and peasant life. They also worked toward weakening the peasant communities' claims to ownership rights over land in favour of warlord supremacy.[67] The *zemene mesafint* was detrimental to development, unsecurity prevailed, the economy declined and the population decreased.[68]

Notes to chapter 2

[1] For classifications of the peoples who speak Kushitic languages in the Ethiopian region see C.F. Beckingham & G.W.B. Huntingford (eds.), *Some Records of Ethiopia 1593–1646* (1954); J. Spencer Trimingham, *Islam in Ethiopia* (1952).

[2] Trimingham, *op. cit.*, pp. 5–9.

[3] Enno Littmann, Reisebericht der Expedition, Topographie und Geschichte Aksums, in *Deutsche Aksum-Expedition,* Band I (1913), p. 41.

[4] *Ibid.*

[5] Trimingham, *op. cit.*, pp. 9 ff.

[6] N.I. Vavilov, The Origin, Variation, Immunity and Breeding of Cultivated Plants, *Chronica Botanica,* Vol. 13, 1949/50. See also J.D. Fage & R.A. Oliver (eds.), *Papers in African Prehistory* (1970).

[7] G.P. Murdock, *Africa: its Peoples and their Culture History* (1959), pp. 181 ff.

[8] Frederick J. Simoons, Some Questions on the Economic Prehistory of Ethiopia, in *Papers in African Prehistory*, pp. 124—129.

[9] Eike Haberland, *Untersuchungen zum Äthiopischen Königtum* (1965), p. 322.

[10] *Deutsche Aksum-Expedition*, Band I, p. 41.

[11] Taddesse Tamrat, *Church and State in Ethiopia 1270—1527* (1972), pp. 16, 21 f.

[12] Sergew Hable Sellassie, *Ancient and Medieval Ethiopian History to 1270* (1972), pp. 71—74.

[13] This can be inferred from stone inscriptions depicting the Aksumite rulers' warlike expeditions. The following extract refers to expeditions made by an Aksumite king in the first century A.D.: "Nachdem ich alle diese Völker, die durch mächtige Berge geschützt sind, in Schlachten, bei denen ich selbst zugegen war, besiegt und unterworfen hatte, schenkte ich ihnen alle ihre Länder gegen Tributleistungen. Aber die meisten Völker unterwarfen sich mir freiwillig und leisteten Tribut." Translated by Littmann in *Deutsche Aksum-Expedition*, Band I, p. 42.

[14] *Ibid.*, pp. 48 f.; Taddesse Tamrat, *op.cit.*, pp. 17 ff.

[15] A.H.M. Jones & Elizabeth Monroe, *A History of Ethiopia* (1970), pp. 31 f.; Taddesse Tamrat, *op.cit.*, p. 26.

[16] German translations of the inscriptions celebrating Aksumite military achievements will be found in *Deutsche Aksum-Expedition, Band IV.*

[17] Sergew Hable Sellassie, *op.cit.*, pp. 137—143.

[18] Taddesse Tamrat, *op.cit.*, p. 23; Cf. Trimingham, *op.cit.*, p. 40.

[19] Taddesse Tamrat, *op.cit.*, pp. 23 ff.; Sergew Hable Sellassie, *op.cit.*, pp. 115—121.

[20] *Deutsche Aksum-Expedition*, Band I, p. 41.

[21] For a detailed account of the monumental architecture of Aksum see *Deutsche Aksum-Expedition*, Band II.

[22] Taddesse Tamrat, *op.cit.*, pp. 31 ff.

[23] *Ibid.*, p. 34; Trimingham, *op.cit.*, p. 44.

[24] Taddesse Tamrat, *op.cit.*, p. 37.

[25] *Ibid.*, pp. 53—60; Sergew Hable Sellassie, *op.cit.*, pp. 239—278.

[26] Taddesse Tamrat, *op.cit.*, pp. 64—68; Sergew Hable Sellassie, *op.cit.*, pp. 283—287.

[27] Taddesse Tamrat, *op.cit.*, p. 119.

[28] For accounts of the evangelization see *ibid.*, pp. 156—205.

[29] *Ibid.*, p. 122.

[30] Trimingham, *op.cit.*, p. 65.

[31] Taddesse Tamrat, *op.cit.*, pp. 80—89.

[32] Trimingham, *op.cit.*, pp. 69—76.

[33] Taddesse Tamrat, *op.cit.*, pp. 297 f.

[34] Manoel de Almeida, The History of High Ethiopia or Abassia, in *Some Records of Ethiopia 1593—1646*, pp. 45—50.

[35] G.W.B. Huntingford, *The Land Charters of Northern Ethiopia* (1965), p. 11.

[36] *Ibid.;* The institution of *rist* has survived to the present day. For an analysis of *rist* rights in land in Godjam province see A. Hoben, Land Tenure and Social Mobility among the Damot Amhara, in *Proceedings of the Third International Conference of Ethiopian Studies* (1970), Vol. III.

[37] Richard Pankhurst, *State and Land in Ethiopian History* (1966), p. 30.

[38] Taddesse Tamrat, *op.cit.*, p. 103.

[39] Bahrey, History of the Galla, in *Some Records of Ethiopia 1593—1646*, p. 126.

[40] The Amharic title of the ruler is *negus*, here translated to king. It is derived from the verb *näggäsä* which means "to collect tribute". See Donald N. Levine, *Wax & Gold*,

Tradition and Innovation in Ethiopian Culture (1972), p. 262.

[41] A collection of one hundred land charters establishing *gult* rights from Aksumite times up to the nineteenth century will be found in Huntingford (1965), *op.cit.*

[42] Haberland (1965), *op.cit.*, pp. 205 ff.; Huntingford (1965), *op.cit.*, pp. 11 ff.; Pankhurst (1966), *op. cit.*, pp. 29 ff.

[43] Taddesse Tamrat, *op. cit.*, p. 101; Huntingford (1965), *op. cit.*, p. 12; Pankhurst (1966), *op. cit.*, p. 49.

[44] *The Glorious Victories of Amda Seyon* (ed. by G.W.B. Huntingford) (1965), p. 18.

[45] Taddesse Tamrat, *op. cit.*, pp. 102 f.

[46] Pankhurst (1966), *op. cit.*, pp. 48, 50, 52 f.

[47] Huntingford (1965), *op. cit.*, pp. 10 f.

[48] *Ibid.*, p. 12.

[49] Trimingham, *op. cit.*, pp. 52 f.

[50] Francisco Alvares, *The Prester John of the Indies*, (ed. by C.F. Beckingham & G.B.W. Huntingford) (1961), pp. 92 f.; Also quoted in Huntingford (1965), *op. cit.*, p. 14 and in Pankhurst (1966), op. cit., pp. 43 f.

[51] Haberland (1965), *op. cit.*, pp. 34–42.

[52] Haberland states that the king's power was limited to choosing between a number of local candidates for the governorship. Haberland (1965), *op.cit.*, pp. 207–210.

[53] Alvares, *op. cit.*, p. 515.

[54] de Almeida, *op. cit.*, pp. 72 f.

[55] Amin (1972), *op. cit.*, p. 107.

[56] Amin (1973), *op. cit.*, p. 10.

[57] According to Amin long distance trade does not compose a mode of production in itself but it can be significant for the transfer of surpluses over boundaries. It has been argued that trade within and through the Abyssinian kingdom was limited in comparision to the trade of Aksumite times, see Haberland (1965), *op.cit.*, p. 205. On the other hand Taddesse Tamrat writes that in the Abyssinian kingdom the royal court was the busiest centre of trade, that the king organized elephant hunting and that the tusks were exported to Egypt, India and Europe by the king's commercial agents, *op.cit.*, p. 88.

In medieval sources quoted by Taddesse Tamrat and Trimingham slaves are mentioned as one major trade item and it seems clear that household slaves were used in Abyssinia.

[58] They are mentioned in Arabic sources quoted in Trimingham, *op.cit.* For a contemporary description of some of the Sidama kingdoms see de Almeida, *op. cit.*, pp. 149–166.

[59] Haberland (1965), *op. cit.*, pp. 318 ff.

[60] For a discussion of the wars in the Ethiopian region between Abyssinia and the Muslim sultanates viewed against the background of the conflict between the Ottoman empire and Portugal over supremacy of the Red Sea and the Indian Ocean see Jean Doresse, *Histoire Sommaire de la Corne Orientale de l'Afrique* (1971), pp. 212–226.

[61] For an account of the invasion see Trimingham, *op. cit.*, pp. 84–93.

[62] Levine, *op. cit.*, p. 20.

[63] A generally accepted explanation is that the Oromos were pushed away from their original homeland somewhere on the plains of the southeastern highlands by the expanding Somalis, see *Some Records of Ethiopia 1593–1646*, pp. lxxi f.; A critical appraisal of the different theories of the original homeland of the Oromo and a discussion of their expansion will be found in Jan Hultin, *The Galla Expansion Reconsidered* (1974).

[64] For accounts of the Oromo expansion see *Some Records of Ethiopia 1593–1646* and Trimingham, *op.cit.*

[65] For an account of the *zemene mesafint* see for example Mordechai Abir, *Ethiopia:*

The Era of the Princes. The Challenge of Islam and the Re-unification of the Christian Empire 1769–1855 (1968).

[66] Trimingham, *op.cit.*, p. 105.

[67] Haberland (1965), *op.cit.*, pp. 214 f.

[68] Abir, *op.cit.*, p. 30.

Chapter 3 The Emergence of the Ethiopian Empire

This chapter focuses on an analysis of the nineteenth century growth of Shoa into an independent kingdom and later into the centre in a new political entity, the Ethiopian empire. The aim of the analysis is to investigate the nature of the relations which were established between the Shoan centre and the conquered parts of the empire. It will be argued that the imposition of new relations of production imprinted deep contradictions on the conquered (southern) regions. The agricultural development projects, which will be scrutinized in chapters 7 to 10, are located in the south. A historical approach to the study of the social structure is essential in order to come to grips with the problems facing the agricultural development projects.

3.1. Shoan Expansion 1700–1847

Prologue

In the heyday of the Abyssinian kingdom Shoa was one of its central regions. The social structure of Shoa showed features similar to the other Abyssinian regions: The basic units of production were peasant families grouped in small villages on kinship basis. Cultivated land and pastures were considered the property of the peasant communities. Individual families were guaranteed the right to use specified plots of land because of their genealogical affinity to the man who first settled in the area, the founding father. Ruling the peasants was a hierarchy of authorities which were local governors performing military and fiscal functions under the Abyssinian king. The governors were authorized to collect tribute from the peasantry. The governors were thus warlords and landlords but not landowners.

The local governors' degree of subordination to the Abyssinian king varied from time to time. The chronicles report that Abyssinian kings at different occasions resided in Shoa. King Amda Seyon (1312–1342) used Tegulat in Shoa as a base from which he took to the field against the Muslim sultanates in the southeast.[1] It is natural to assume that the king at such times had the local governors under direct supervision.

Ahmad Gragn's *jihad* in the sixteenth century and the ensuing penetration of the highlands by Oromo tribes weakened the actual influence of the kings. Shoa was virtually cut off from the other Abyssinian regions by Oromo incursions. Even within Shoa the Christian Amharic speaking population was confined to a small number of isolated localities under continuous Oromo pressure.[2]

Only after a century and a half were the Christian, Amharic speaking Shoans capable of taking the political and military offensive. In the turn of the seventeenth century one local ruling house initiated the slow process of subduing the neighbouring Amharic as well as Oromo areas. The expansive ruling house of Shoa originated with Nagasse Kristos (died 1703), who claimed to be a descendant of the Abyssinian king Lebna Dengel.[3] Nagasse and his successors forcefully imposed their supremacy over a growing number of Amhara and Oromo districts and demanded tribute from them.[4] They maintained sporadic contacts with the Abyssinian king in Gondar, some of them travelling there in order to receive official sanction for their position as rulers of Shoa. The king had, however, little actual control over the situation in Shoa.[5]

The descendants of Nagasse, which became known as the Manze house continued their expansion during the eighteenth century.

In the early nineteenth century Shoa emerged as an independent political entity under the Manze rulers, to which the original Amhara districts as well as a growing number of Oromo districts paid tribute. Sahla Selassie (1813–1847) claimed the title king (*negus*) of Shoa and thus elevated the Manze house to a ruling dynasty.[6]

Political Centralization

For more than 30 years Sahla Selassie embodied the central political authority in Shoa. The local authorities consisted of powerful families with ancestry in their districts. Sahla Selassie's policy aimed at centralization in the sense of control over appointments and removals of local governors. He sought to thwart the principle of hereditary rule on the local level. A European traveller commented on the king's intentions:

> "Most probably the judicious monarch only waits for an opportunity of doing away with all hereditary governments in the kingdom".[7]

During the latter part of Sahla Selassie's government the king appears to have gained control over the local authorities. According to Abir, Sahla Selassie was an absolute monarch in the districts which composed the core of Shoa. Here the local authorities were either persons directly appointed by the king or hereditary chiefs. Even in the case of hereditary chiefs the king decided which son was to succeed his father.[8] The king further countered the power of the hereditary chiefs by appointing in every district an official, called *mislene*, who was responsible for tax collection and mobilized the peasants for work-days on royal agricultural land. Abir assumes that the administrative power was in the hands of *mislenes* especially since the hereditary chiefs had to be away from home for long periods of time accompanying the king on warlike expeditions.[9] The tactics of Sahla Selassie were thus to double the number of local authorities and to transfer administrative power to officials directly appointed by him.

The bulk of the population was peasants who performed various services to the local and central governments. Domestic slaves, originating from the Oromo and Sidama areas, were commonly held and were either taken in campaigns or bought from slave traders, "from the governor to the humblest peasant every house possesses slaves of both sexes, in proportion to the wealth of the proprietor".[10]

Territorial Expansion

Against the Oromo groupings the Manze dynasty pursued a combined policy of conquests and alliances. The Oromo tribes surrounding Shoa were not politically united, hence it was possible for the Shoan rulers to make an alliance with one tribe in order to subjugate another. At first Shoan influence over the tribes was shaky and alliances were refuted by Oromo chiefs when the Shoan ruler died. The first task of a successor was therefore to wage war against the rebelling tribes in order to restore Shoan sovereignty and to bring additional tribes into vassalage.[11]

The basis of Shoan expansion was military strength. The Shoan rulers were above all warlords. Of great importance was the fact that they realized very early the decisive role of fire-arms. A sporadic import of weapons into Shoa had occurred before the time of Sahla Selassie but it was he who systematized the import by making agreements with French, Italian and British traders in arms.[12] In general the king showed interest in making use of foreign technical innovations.[13]

But military strength was not the sole decisive factor in Shoan expansion. Personal diplomacy and alliances also played important roles. The original Amharic Christian population in Shoa was by far outnumbered by the surrounding Oromos.[14] The Shoan rulers therefore found alliances advantageous. Oromo chiefs who responded were rewarded with Abyssinian titles, Christian Amharic names, wives from the Shoan court etc.[15] The alliances were military as well as political. Oromo warriors joined the Shoan army in military campaigns while Oromo chiefs became local governors under the Shoan king. It can be argued that all parties benefitted from such alliances; the Shoan king expanded his territory, the Oromo chiefs joined the privileged groups who benefitted from the expansion and the Oromo warriors had greater opportunities to pillage when they operated as units of the strong Shoan army. The object of yearly raids was Oromo tribes which had not joined the alliances.

It can be argued that the direction of Shoan expansion was conditioned by economic motives. Within Shoa no valuable items of trade were produced.[16] The surpluses appropriated from the peasantry were consumed by the authorities. But Shoan rulers needed to gain control of resources which could be exchanged on the international market for fire-arms; a necessary precondition for Shoan supremacy and further expansion. By moving south they gained control over strategically situated markets along the major trade route between Red Sea ports and the Oromo and Sidama kingdoms further southwest of Shoa.[17] Until Sahla Selassie's reign the Shoans had to confine themselves to controlling some of the trade routes. In the latter part of the nineteenth century, however, expansion was directed towards the rich production areas themselves.

Shoan expansion was temporarily halted in the mid nineteenth century. The warlord Kasa, a military genius from Godjam, gained control of the political situation in most of the regions which had comprised the old Abyssinian kingdom. Kasa defeated the other warlords, including king Haile Malakot of Shoa, and had himself crowned Emperor (*negusa negast*) in 1855 under the name Tewodros.[18] For a short period Tewodros held Tigre, Beghemder, Lasta, Godjam and Shoa under his sway. But repeated uprisings exhausted his military power and when a British military force joined the rebels the Emperor was defeated in 1868 and committed suicide in the face of defeat.[19]

After the death of Tewodros the British withdrew.* The Tigrean warlord Kasai, who had helped the British and was supplied with many fire-arms by them, commanded now the strongest military force in the Abyssinian regions. He proclaimed himself Emperor under the name Yohannes.[20]

Yohannes ruled until 1889 but was most of the time involved in military and diplomatic operations on the northern and western frontiers. During this period the new Shoan king Menilek, grandson of Sahla Selassie, was free to consolidate his power in Shoa and to renew the drive to the south.

In the following sections Shoan expansion will be analyzed. An overview of the social structure of the societies which became objects of Shoan expansion will be given first however.

* Their aim had been fulfilled: to release a British consul whom Tewodros had imprisoned.

3.2. A Survey of the Southern Areas before their Incorporation into Menilek's Empire*

One can roughly classify the societies located in the south as either pastoralist or agricultural. Rift valley, the southeastern highlands as well as their adjacent plains and lowlands were inhabited by tribes with a pastoral economy; in the Awash valley the Afar lived, in Ogaden the Somali lived and in the southeastern highlands and their adjacent plains were different Oromo groups. The pastoral Oromos were organized in clans whose internal government was performed through a complex organization of age groups and age sets called the *gada* system.[21]

The Sidama societies of the southwestern highlands stood in sharp contrast to the decentralized political systems of the pastoralists.[22] Their economy was agricultural (which included livestock breeding) and political centralization was high. The most famous Sidama kingdom was Kafa. Kafa was ruled by a king and a royal council, while the population was grouped in clans. Regional government was in the hands of clan leaders, the foremost of which belonged to the royal council.[23] Other Sidama kingdoms were Wolamo, Kambatta, Janjero. They were all characterized by a common social structure which included a sacred king whose behaviour was surrounded by rituals, a class of nobles functioning as royal advisors and regional rulers, a peasantry, craftsmen and slaves. The Sidama kingdoms were located in very fertile areas and due to the relatively advanced level of the productive forces (irrigation and terracing were used in many places) large surpluses could be appropriated by the rulers. In addition, trade contributed significantly to the wealth of the Sidama kingdoms. The main trade routes ran to the Red Sea ports, one via Harar and one via Godjam and Tigre. One route also ran west into Sudan. The main items of export were slaves, cattle, honey, wax, and to some extent coffee.[24] These Sidama kingdoms had earlier paid tribute to the Abyssinian king, but Oromo offensives had put an end to Abyssinian influence.

* The areas in question are actually located west, southwest, south, southeast, east and even northeast of nineteenth century Shoa (see map 4 and 5). For the sake of convenience they will be referred to as the south, the southern regions, the provinces in the south etc. These areas have in common the fact that they were reduced to political and economic dependency on Shoa. The areas to which I collectively refer as the southern regions correspond roughly to the following administrative units in the present day Ethiopian empire: Wollega province, Shoa province south and west of Addis Ababa, Wollo province except Wag, Lasta and Wadla Delanta *awradjas*, Arussi province, Hararge province, Kafa province, Illubabor province, Gemu Gofa province, Sidamo province and Bale province. See map 6.

To the north and west of the Sidama kingdoms, around the Gibe river, there were a number of Oromo kingdoms. They had been established in the seventeenth and eighteenth centuries when pastoral Oromo tribes swept into the Gibe region and conquered the Sidama kingdoms which already existed there. The victorious Oromo settled in the region and adopted agricultural practices and monarchial institutions from the vanquished Sidama population. The Oromo kingdoms in the nineteenth century were Jimma, Limmu-Enarea, Gera, Guma and Gomma.[25] They were hereditary monarchies. Under the king there was an aristocratic group of regional authorities, a peasantry and an apparently quite large slave population which constitued a major trade asset. Agricultural production was centred on cereals and much live-stock was kept. Similar to the Sidama kingdoms, trade, controlled by the king, played an important role in the economic life of the kingdoms. Besides slaves, civet, wax, honey, ivory and coffee were traded.[26]

Along with the trade came Muslim influences. In the mid nineteenth century Islam was adopted and started to penetrate from the royal court to large sections of the population.[27]

In the latter part of the nineteenth century Jimma was the most powerful among the Oromo kingdoms. It was ruled by Abba Jifar, whose government was firmly centralized. The kingdom was divided into 60 districts, the regional governors being appointed by the king and paid from the central treasury. The governors were mainly responsible for law and order and the peasantry paid taxes directly to royal collectors.[28]

Land in Jimma could be bought and sold. It was possessed by private persons, the king being the largest landowner. In an attempt to reconstruct the land tenure system in Jimma before its incorporation with Menilek's empire, Lewis writes:

> "In Jimma a rich landowner may have one or more great parcels of land (*yebbo*) on which he and his followers and tenants reside. Before 1932, a rich landowner would have slaves as well. The tenants (*kubsisa*) owe the landlord a fixed amount of labour or may pay rents in currency or kind. The *abba lafa* ("lord of the land") appoints stewards to see to the performance of the tenants' obligations. In the days of the monarchy, if a landowner was a government official... he would spend a great deal of time at the palace away from his lands leaving them to the care of his stewards. In addition to consolidated land holdings, men may have properties in different parts of the country. The people who live on such absentee lands (*gufo*) pay rents to the owner or his representative."[29]

The private ownership of land originated when Oromo war leaders, the

Map 4. SOME POLITICAL AND TRIBAL GROUPINGS IN EARLY 19TH CENTURY ETHIOPIA

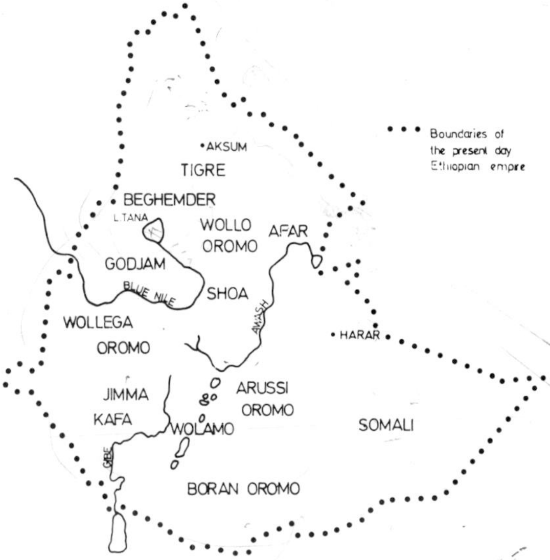

Map 5. MAIN DIRECTIONS OF MENILEK'S CONQUESTS IN THE 1880's AND 1890's

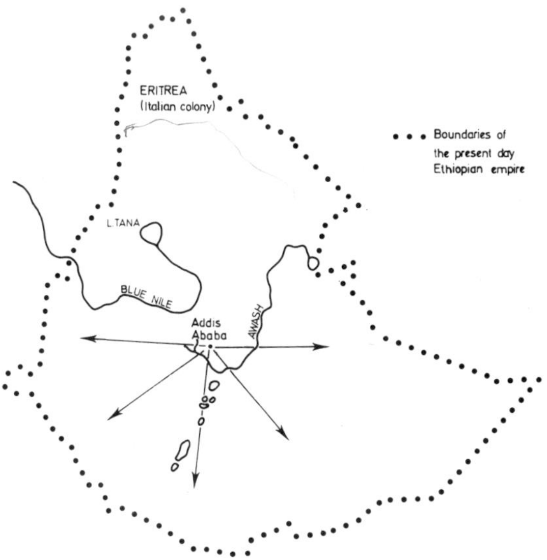

Map 6. PROVINCES OF THE PRESENT DAY ETHIOPIAN EMPIRE

Source: Central Statistical Office.

digo, conquered Jimma and appropriated land for themselves. It was later intensified by a flourishing trade which provided the privileged groups with money.[30] Consequently, the status of the peasantry was progressively reduced to that of tenants.

Despite apparent differences between the Sidama and Oromo kingdoms, such as language and religion, their social structure was quite similar. I suggest that they can all be characterized by a dominant tributary mode of production. Peasant communities and a state structure were the two main juxtaposited elements, and surpluses from the subordinated peasantry was the source of the kingdom's wealth. It is evident that slavery also was of great economic significance, slaves being used both in agricultural production and as a trade asset. The rulers' wealth was augmented by the fact that the tributes they exacted from the population had a high market value among the merchants from all over the Horn of Africa, so that the rulers had great incomes from trade. It is interesting to note that, in Jimma at least, the tributary mode of production developed into feudalism in the nineteenth century and that this appears to have been conditioned by increased commercialization.

The region between Gibe and the Blue Nile, today the province of Wollega and the western part of the province of Shoa, was also Oromo land. The Oromo tribes who reached this region did not become fully settled as did their neighbours further south. They maintained a high emphasis on pastoralism and were politically decentralized. Their foremost political institution was the *gada* system. In the course of the nineteenth century, however, they became more stationary and shifted the emphasis of their economy to agriculture, although animal husbandry remained important.[31]

It seems that the decentralized government was linked to their pastoral economy, being weakened when the tribes settled and surpluses from agricultural production invited a more unequal distribution of wealth.

> "In the earlier 19th century important social and institutional changes took place. The traditional *gada* system of social, political and religious organization began to give way to systems of individual rule and authority, perhaps as the result of increasing class divisions based on land. By 1880 much of the [Quelem] *awraja* was under the rule of Jote' Tulu, whose state had been greatly expanded from its primitive beginnings under his forefathers".[32]

The Oromo societies were in the process of developing into small kingdoms, the process had progressed furthest in Quelem under the reign of Jote' Tulu and in Nakamte under the reign of the Morodda family.[33]

In the valley of the Blue Nile and its tributaries lived 'Prenilotic' tribes referred to as Bani Shangul. They subsisted from hunting, fishing and agriculture.[34]

3.3. General Reasons for Conquest

What follows is a summary of why the above described areas became subjected to military conquests at the end of the nineteenth century.

Economic Importance

Internationally sought raw materials originated in the Oromo, Sidama and Bani Shangul areas. The following products were found:

Gold

The bulk of the gold extracted in the Horn of Africa came from the river valleys in Wollega where it was buddled by the local inhabitants and sold to merchant caravans.[35]

Ivory

Elephants were found in abundance here, while they were becoming rare in the Abyssinian regions during the second part of the nineteenth century due to intensive hunting.[36]

Musk
The civet-cat was common. It was kept in cage and it produced an ingredient for perfume which was highly valued in the Orient.[37]

Slaves
The bulk of the slaves which were sold to the Orient from the Ethiopian region came from the Sidama and Oromo areas. Most slaves were prisoners taken in raids between the different kingdoms. In Kafa and Jimma the kings had a monopoly on the sale of slaves to foreign traders, but the traders also raided unprotected small villages along the caravan route. According to some estimates 25,000 slaves yearly were exported from the Oromo and Sidama areas in the latter part of the nineteenth century. The slaves for the Muslim world were exported through Sudan or through Red Sea ports.[38] The most coveted slaves were young Oromo girls who commanded a high price in the harems of the Orient.[39]

Coffee
The southwestern highlands is probably the area where coffee originated. The coffee bush grew wild in the forests. The berries were picked by the local population and used for home consumption and as a tribute to the rulers, who in turn sold it to merchants. Only quite small quantities were sold, however.*[40]

Beside these raw materials the Oromo and Sidama areas were known in Abyssinia for their fertility and abundance of cattle and agricultural products. This must be contrasted with the situation in the Abyssinian regions where crop failures often occurred and a high degree of deforestation had been realized by the second half of the nineteenth century.[41]

Political Weakness
The lack of political strength and unity increased the attractiveness of these areas. Kafa, Jimma and Wolamo notwithstanding, the Oromo and Sidama societies were 'lilliput' kingdoms which often fought each other. The political unity of the pastoralist societies, with loose and shifting alliances between different clans, was even weaker. The wars between these kingdoms and tribes were mostly skirmishes. The warriors fought with

* Commercialization on a large scale was not begun until after world war II.

spears on horseback and were not acquainted with fire-arms.[42] The 'army' which king Abba Boka of Jimma sent against the neighbouring kingdom of Guma in the mid nineteenth century consisted of 44 horsemen. Huntingford comments:

> "This small force of forty-four is not merely fable: it indicates the true nature of this kind of warfare, which was carried on by quite small bodies of men on each side."[43]

Later in the century when Abyssinian warlords could increasingly supply their armies with fire-arms, it became apparent that it was possible to conquer the areas in the south. An Abyssinian scramble for these areas developed. Both the ruler of Godjam, king Takla Haimanot, and the ruler of Shoa, king Menilek, sent out military expeditions. The rivalry between the two was decided in 1882 when Menilek set out against Takla Haimanot in order to prevent him from conquering Kafa. The rivals encountered each other in Wollega where the king of Godjam was decisively defeated and taken prisoner by Menilek.[44] Menilek had thereby obtained sole rights over military expeditions to the south. The only one who could have challenged him was the ruler of Tigre, Emperor Yohannes, who was formally Menilek's superior. But Yohannes was until his death in 1889 busy continuously countering attacks by the Egyptians, Sudanese dervishes and Italians in the northern and western Abyssinian regions and could not check Menilek's activities in the south.[45]

3.4. The Conquests

In the early 1880's Menilek embarked on systematic campaigns to conquer the areas south, southwest and southeast of Shoa. During the preceeding decade he had built up a stable platform from which to operate. Rivals to the Shoan throne had been eliminated and Emperor Yohannes had recognized him as king of Shoa. Menilek made a number of political alliances with powerful warlords and he often used marriage as an instrument.[46] An important alliance was made between Menilek and Gobana Daci, an independent Oromo warlord in Shoa. In the turbulent years of Emperor Tewodros, when no one held undisputed power in Shoa, Gobana chose to ally himself with the young Menilek. Gobana submitted to Menilek, who made him a high ranking officer in his army. In command of a large body of heavily armed Oromo horsemen Gobana performed many of the conquests for his lord Menilek until his death in 1889.[47]

Although the Shoan military forces in extensive campaigns included a large number of peasants mobilized by their lords on the order of Menilek, they also included increasing numbers of mercenaries; Tigreans and Amharas from the Abyssinian regions and Oromos from Shoa.[48] Menilek gave high priority to the import of fire-arms and was successful in this since France and Italy were willing to supply him with weapons. Italy hoped, up till 1889, to form an alliance with Menilek against Emperor Yohannes, while France saw Menilek as a potential ally against the British in East Africa.[49]

Menilek's residence was situated on the mountain Entotto (immediately north of present day Addis Ababa). During the 1870's Menilek, using a mixture of persuasion and military threat, had forced the Gulalle Oromo which lived on the plains below Entotto to submit to his authority.[50] He thereby received a strategical staging zone and a further increase in Oromo cavalry for coming expeditions.

The first series of Shoan military expeditions on a large scale was directed against the Arussi highlands, south of Shoa. From 1882 to 1886 Menilek sent out eight expeditions against the pastoral Arussi Oromo. Facing the Shoan offensive the Arussi clans united and presented fierce resistance.[51] In the long run Shoan fire-power was the decisive factor. The expedition of 1886 developed from a battle into a slaughter:

> "The Shoan army passed the Tehokie range killing and mutilating . . . the Arussi children, women and elders; villages and farms were burnt . . . the mutilation continued up to the banks of the Webi and the borders of Sidamo."[52]

Arussi resistance was broken, some clans gave up, while others moved away to the Bale highlands. Menilek secured his control over Arussi by leaving a garrison, a *katama*, behind.[53] Since Arussi had been deprived of a great deal of its original population (due to the massacres and the subsequent disappearance of some clans) there occurred a migration of people into Arussi after the conquest. Menilek granted large tracts of land to his followers, who in turn invited peasants from Shoa to settle as cultivators on their grants.[54]

When returning from Arussi Menilek sent an expedition east to the city of Harar. This once so powerful trade centre was at the time in a state of decline and could not offer much resistance. The Emir of Harar was defeated and the emirate incorporated into Menilek's empire.[55]

In the 1880's Menilek also increased his influence in the highlands southwest of Shoa. In 1882 king Abba Jifar of Jimma submitted peacefully to Shoan supremacy. Relations between Shoa and Jimma were established so that Jimma kept its inner autonomy (government by Abba Jifar, Islam) but was obligated to pay tribute to Shoa. A yearly tribute was paid and Abba Jifar pledged to give military assistance to Menilek on his subsequent war expeditions. The other Oromo kingdoms in the Gibe region were forcefully subdued and put under Shoan military government.[56]

The Oromo communities in Wollega were incorporated in a way similar to Jimma. Menilek's general Gobana was sent to Wollega where some chiefs resisted and were defeated, while others were persuaded to peacefully accept Shoan supremacy.[57] Many Oromo chiefs made the best out of a threatening situation. One of the most powerful chiefs was Joté Tulu who ruled the Quelem area:

> "Joté, clearly anticipating the results of resistance, bowed for the expanding Shoan state and submitted to *Negus* Menilek. Until 1918, the year of Joté's death, Qelem enjoyed a semi-autonomous status. Thereafter, it was absorbed into the more normal pattern of rule from Addis Ababa through governors foreign to the area".[58]

A similar pattern was repeated in other Oromo communities in Wollega. In return for their inner autonomy the chiefs, now nominally Shoan governors under Gobana or directly under Menilek, paid an annual tribute in gold and ivory to Addis Ababa.[59]

In the 1890's the drive south was accelerated. The earlier conquests had opened the way for Menilek's armies to the rich highlands further to the south. In the early 1890's Sidamo, Bale and Ogaden were conquered. Their fate was military occupation. *Katamas* were established and the occupant Shoan military command functioned as Menilek's regional government.[60] In 1894 the Sidama kingdoms Wolamo and Janjero were defeated. Wolamo gave battle under its king Tona, but resistance was once again broken by the superior Shoan fire-power. As in Arussi the battle turned into a massacre, which decisively wiped out Wolamo resistance. Menilek appointed the defeated king Tona as his regional governor in Wolamo, but the real power was vested in the Shoan military commander who established a *katama* in Wolamo-Soddo.[61]

During the two following years Menilek was totally occupied with the Italians in Eritrea and little activity took place on the southern front. But after the victory over the Italian army in 1896 military pressure in the south was renewed. In 1897 Menilek decided to conquer the powerful and rich Sidama kingdom of Kafa. As early as in 1882 Kafa had agreed to pay a yearly tribute to Menilek and at that time the Shoan king could hardly have exacted further concessions from Kafa. But during the following decade Menilek's conquests had resulted in a virtual encirclement of Kafa. The Shoan armies could now reach the wooden palisade which demarcated the frontier of Kafa, without having to pass through enemy territory.[62]

A military coalition consisting of the Shoan army under general Walda Giyorgis and an army from Jimma under Abba Jifar waged war against Kafa in nine months.[63] In open battles the assailants were victorious while the defenders found guerilla tactics more successful.

> "The Amhara had guns, we had only spears. We hid behind the trees and speared them. But they had many guns and we had few spears".[64]

At last the resistance was broken and a massacre followed.

> "The conquerors now committed a series of atrocities. They burnt the native residences, took away their wives, children and some cattle and set crops ablaze."[65]

Kafa was reduced to a province under Shoan military rule.

During the last three years of the century Menilek's armies proceeded to advance from the highlands and occupied the Bani Shangul lowlands in the west and south west, the Borana plains in the south, and the Aussa sultanate in the lower Awash valley in the east. Shoan soldiers were posted in *katamas* everywhere except in Aussa, where the sultan was allowed to remain in power in exchange for yearly tributes to Menilek.[66] These semi-desert and jungle areas formed buffer zones between the emerging Ethiopian empire and the surrounding European colonies.[67]

Meanwhile, Menilek's authority had also expanded northwards. When the supreme Abyssinian warlord, Emperor Yohannes, died in 1889, Menilek was recognized as the new Emperor by the warlords in the Abyssinian regions.[68] His empire, the centre of which was Addis Ababa, founded in 1886 below Entotto mountain, thus extended over a larger area than any Ethiopian ruler had held before. The possibilities of Shoan territorial expansion were thereby exhausted. In all directions it encountered the simultaneously advancing European imperialism. The victory over the Italians in 1896 forced the European powers to recognize the Ethiopian empire, although their colonial aspirations over Ethiopia did not disappear.[69]

At the same time as Menilek resisted European imperialism he participated in the scramble for Africa. His territorial ambitions are reflected in a letter which he sent in 1891 to the European heads of state and where he laid claim to all land east of Fashoda and north of Lake Rudolf.[70]

3.5. Some Specific Reasons for the Conquests

Any attempt to interpret Menilek's drive to the south must dwell at length on economic factors. Without aspiring to analyze the situation in depth, I will elaborate on some points of importance.

Menilek's expansion followed a pattern of crude dynamism. Since the Shoan economy was not built upon monetary exchange Menilek could not pay his soldiers in cash. The only immediate reward he could offer them was plundering. Descriptions of the military expeditions tell of the destruction; villages were sacked, crops burnt, people killed and mutilated. When the expeditions were terminated the bulk of the army marched off with slaves and cattle. The conquered areas were reduced to poverty. As soon as the soldiers had consumed or sold their booty, they returned to Shoa. Addis Ababa, however, could not feed a standing army. Consequently, Menilek could not afford to have a growing number of mercenaries and semi-professional soldiers quartered in Addis Ababa for long periods of time. Their maintenance required that they be kept on the war-path and continuously acquire new areas to plunder.

In the early 1890's an economic crisis in Shoa made access to the resources of the south a prerequisite for survival. In 1888–89 a rinderpest spread into Shoa from the north and proceeded further south, killing off nearly all the cattle. In Shoa and the northern regions the rains also failed causing subnormal harvests. The lack of work oxen rendered preparation of the fields for the coming harvest utterly difficult.[71] Menilek anticipated the catastrophe:

> "Hommes du Choa, réunissez-vous tous dans vos églises et poussez vers Dieu le cri d'*Egzio*, car vous savez bien que, si les boeufs disparaissent, il n'y aura plus de grain, et que, s'il n'y a plus de grain, il n'y aura plus d'hommes".[72]

The ensuing famine decimated large sections of the population.[73] The 1890–1894 Shoan war expeditions against different Sidama and Oromo kingdoms which were less famine-stricken must have been directly conditioned by the famine. The Shoan army had to conquer food supplies from distant areas in order to survive. When the crisis reached its peak the poorest sections of the population in Shoa were dead. "Everybody felt the lack of manpower and servants so that Menilek wanted to re-supply the manpower by slaves".[74] This was the immediate reason for the expedition against Wolamo in 1894, while the result of the successful expedition was a heavy depletion of Wolamo's resources. The Shoan army marched off with a huge quantity of slaves and cattle. Especially important were the thousands of head of cattle which were to repopulate Shoa with livestock.[75]

If the broader perspective of international politics in the Horn of Africa is considered, it is obvious that Menilek's drive south was related to the European scramble for Africa in general and to Italian pressure on the Abyssinian regions in particular. An analysis of these international relations is beyond the scope of this book. It must suffice here to state that Italy had colonial ambitions in the Horn of Africa. In 1885 Italian troops occupied Massawa and proceeded to the Tigrean plateau, which, north of Mareb river, became the Italian colony of Eritrea. When Menilek succeeded Yohannes in 1889 as Emperor he concluded the Uccialli treaty with Italy.[76] The treaty was to regulate the relations between the two powers. The Italian interpretation of the treaty was, however, that Ethiopia had accepted the status of a protectorate under Italy.[77] Menilek opposed such an interpretation (which found no support in its Amharic version) since it threatened Shoan expansion from an independent centre. Menilek remained passive for a few years, during which time he increased the import of arms from France and Russia. He then officially denounced the treaty in 1893.[78] After that, there was no other alternative than a military confrontation.

All of Menilek's more important activities during the period 1890–96 can be interpreted as being aimed towards mobilization of resources in the face of an inevitable clash with Italy. Activities on the southern front were related to the Italian question in the sense that an opening up of these areas gave Menilek access to resources which could be used to strengthen his army (gold and ivory could be exchanged for weapons, slaves and animals were needed for the large expedition against the Italians). These resources could also be used as potential rewards after a victory.

The course of events of the Ethiopian-Italian war can be summarized very shortly. The decisive battle was fought in 1896, at Aduwa in Tigre where the Italian forces were thoroughly beaten and the Italian threat was postponed.

In the period following Menilek's victory, the problem of rewarding the soldiers and officers had to be solved. The army Menilek commanded at Aduwa was huge. The Ethiopian troops at Aduwa were reported to have amounted to some 100,000 men under arms.[79] Many of these must have been peasants, mobilized for this specific war by their lords, and they awaited no other reward than the booty of the battlefield. But the mercenaries and the warlords expected to be rewarded.

The military operations of the 1880's and 1890's had greatly enlarged the group of professional and semi-professional soldiers who, if discontented, could be dangerous for the Emperor. Pankhurst hints at this when he writes:

> "Menilek, particularly in later years, was faced with the existence of large numbers of unpaid but virtually professional troops, whose numbers had swollen during the various campaigns against the Egyptians, Dervishes and Italians".[80]

After the victory the problem of reward was thus acute. The troops were hungry. Menilek's resources in hard currency were limited and the Abyssinian regions were more exhausted than probably ever before in the aftermath of the famine. The situation had deteriorated further through the deployment of the armies throughout Abyssinia.

Thus the problem of rewarding the troops could only be solved by granting land and labourers to the victorious soldiers in the newly conquered south. Here were the most fertile agricultural regions in the empire.

3.6. Mechanisms of Surplus Appropriation

The Gabar System

The expression *gabar* came under Menilek's time to signify a peasant from whom civilian or military officials were empowered to exact labour and agricultural products. This form of servitude will in the following text be referred to as the *gabar* system.

The conquests were concluded at the turn of the century. A number of uprisings had been brought under control by the *katama* troops. The population in the south was subdued.

The legitimization for the introduction of the *gabar* system was that Menilek, following the Abyssinian and Shoan royal tradition, proclaimed himself overlord of all conquered land.[81] Menilek applied this principle with sword in hand and utilized it as a reward system. The procedure was the following. First the conquered land was proclaimed the property of the Imperial Crown. Then Menilek granted this land to his followers and other persons who demanded remuneration in return for their services to the Crown. In general, one-third to one-fourth of the land in each conquered region was granted to kinglets, chiefs or other leaders among the subdued peoples who showed interest in cooperating with Menilek. They were recognized as local representatives of the Imperial power and were called *balabbat*. The remaining two-thirds to three-fourths were partly kept by the Crown and partly granted to Shoan military men who were stationed in these areas.[82]

It should be noted that Menilek did not grant ownership of the land, only the right to exact tributes from the peasants living in the granted areas. These grants were valid only as long as the beneficiary was in the active service of the Crown. The number of peasants who were 'distributed' varied with the rank of the beneficiaries. Perham mentions that common soldiers received 2 to 5 peasants who worked for them, warrant officers received 7 to 10 peasants, higher officers 30 to 80 peasants and provincial governors several hundred peasants.[83] These peasants were the *gabars* of their lords. The obligations of the *gabars* were not fixed and thus were often decided by the arbitrary will of the beneficiary. It seems that the *gabars* under common soldiers were subject to the most severe exploitation. The soldiers were themselves poor and utilized their privileges for maximum exploitation of the peasant families which at the time were at their disposal. The exploitation of the *gabars* belonging to high officials was generally somewhat more moderate, probably because these lords were already wealthy and had no interest in driving their *gabars* to a state of despair. Concerning the situation in Kafa F. deHalpert reported that,

> "The people preferred working for these great landlords, for they were thereby protected from the extortions of the petty officials and soldiers. Owing to the diminution of the population there was competition for labour and there was a tendency for *gabars* to escape to the estates of the great landlords, who were glad to keep them as tenants and often prevented them from being reclaimed".[84]

Pankhurst's summary of the obligations imposed on *gabars* belonging to high officials indicates that between one-third and one-half of their working time was allocated for the benefit of their lords.[85] Their primary responsibility was the production of agricultural products. The lords demanded from one-third to one-half of the harvest. In addition they had to provide chicken, eggs, sheep and beer for their lords' banquets on important holidays. They could also be placed on construction projects for 'the public good' (roads) or for private interests (houses for their masters).[86]

The personal security of the *gabars* was non-existent. In spite of being officially banned, slave trade continued. Menilek's regional governors sometimes acted as suppliers of slaves to the merchants. The most horrible examples are reported from Kafa, where one provincial governor imposed such severe taxes that the population could not pay. The governor thus forced the peasants to give away their wives and children in lieu of paying the taxes. The women and children were then sold to traders.[87] It is reported that slave caravans continued to move from the southern provinces to the Red Sea until 1935.[88]

The tough and brutal style of the government of the southern provinces was partly due to the ethnic and cultural differences between the conquerors and conquered. The former called themselves Amharas. This was a truth with modifications however. Among Menilek's followers there were a number of persons from the Abyssinian regions and these were Amharas as well as Tigreans. But the bulk of the conquerors were Shoans and it has been pointed out that the Oromo element in Shoa was considerable. But the Shoan Oromos, and especially their chiefs, adopted the official Amharic culture of the Manze dynasty. They learned the Amharic language, changed their names to Amharic names and joined the Orthodox Church. Haberland considers the amharized Shoan Oromos the most loyal followers of the Manze dynasty, from which many of them have received privileges and high official positions.[89]

The conquered peoples were Sidamas and non-Shoan Oromos, who were considered inferior by their conquerors.

The consequenses of the *gabar* system for peasant social life in the affected areas have not been fully documented. But the scanty information which is available indicates that there were devastating effects. The situation seems to have been worst in Kafa, where the exploitation was equal to pure pillage. Consequently many *gabars* preferred to give up agriculture altogether and flee to the forests in order to escape their tormentors. The amount of agricultural land consequently decreased and wilderness reclaimed much of the area.[90] The *gabar* system in Borana, adapted to the pastoral community, made many of the Boran Oromo clans escape over the border to British East Africa.[91]

Land Tenure

Menilek reserved large areas of land for the Crown. Imperial estates were established which served to supply the palace in Addis Ababa with food stuffs. There were different kinds of such estates: *mad bet* and *hudad* were Crown land where grain and pepper was cultivated, and *wärä gän* was Crown land where cattle was bred. Thus the Imperial Crown emerged as a landowner of great magnitude in southern Ethiopia. The agricultural work on such estates was performed by the peasants in the surrounding villages or by slaves of the palace. The products from these estates were consumed in Menilek's palace in Addis Ababa where banquets were held for large numbers of soldiers and officers.[92]

The privileges which the officials received in relation to the peasantry in the southern provinces and which constituted their salary, were institutionalized in new forms of land tenure. The new tenures were modifications of the old *gult* privileges of the Abyssinian kingdom. The three most important tenures were *maderia, rist-gult* and *siso.*

Maderia was given to the lower military command and to common soldiers. This tenure, which still existed in the 1960's, is officially defined as follows:

> "Maderia land is Government land granted to Government employees in place of salary or as a pension for the period of office or for life. A maderia holder must pay asrat [tithe] health tax and education tax on his holding. He may collect rentals in produce from persons settled on his holding. Maderia land essentially belongs to the Government which can transfer it to another person when the maderia holder is convicted of crime or transferred to another job. He may not, of course, dispose of his holding by gift or sale and his holding is not heritable".[93]

Mahteme Sellassie mentions that a simple soldier could get a maximum of 3 *gashas* (120 hectares) of *maderia* land. High ranking officers received larger areas.[94]

The land granted to high ranking officers was of the *gult* type. It was clearly connected with the exercise of their official duties.

> "Les provinces plus éloignées d'Addis-abéba furent données par Ménélik à ses lieutenants généraux, assez souvent à ceux qui les avaient conquises, mais à titre essentiellement temporaire . . . Le *goultégna* ne devient pas *ipso facto* propriétaire des terres comprises dans son fief, mais il a l'usufruit de certaines et, sur les autres, il perçoit les impôts, exerce les droits de réquisition et de corvée, à charge par lui de rendre la justice et de fournir en temps d'expédition, un contingent déterminé".[95]

To his favourites Menilek granted a special tenure called *rist-gult*. In essence, this was *gult* rights which were inherited by the grantee's children. Mahteme Sellassie writes:

> "Rist-Goult was land given for services rendered. Its beneficiary became its proprietor and could transmit it to his successors. He profited from the entire revenue and collected the taxes for his own use. His heirs to gain possession of "rist-goult" had to obtain the authorisation of the Government and to present a mule, after which the land was transferred to their name".[96]

Land given to *balabbats* was called *siso*, its content was similar to that of *rist-gult*.[97]

Rist-gult thus represents a step in the direction of the privatization of land ownership. Menilek adopted further measures in this same direction. As early as 1891 he issued a proclamation which strengthened the land holding of private persons in the southern provinces. Pankhurst relates:

"Menilek ... issued an edict establishing hereditary ownership of land in the area between Wayt and the Awaš. The decree declared that all Galla land known as *gaša märet* (shield land) the holders of which were supposed to furnish a man armed with a shield, should henceforward be considered as *rest*, or hereditary property."[98]

Those who benefitted from this proclamation were the *balabbats* and military officers who were already established in these areas. Gabra Selase's chronicle relates that the proclamation filled the soldiers and officers with great joy.[99] How the *gabars* reacted is not mentioned.

One more proclamation was issued in 1907 by Menilek's newly appointed Government. Again, Pankhurst relates:

"The ministers further proclaimed that all land inherited from father or mother or purchased with money was to be considered as *rest'*... The decree also laid down that lands belonging to a person who died without children should pass to the deceased's father or mother and not to the local chief, or *šeqa šum*, who had hitherto taken them as the representative of the Government ... The right of a proprietor to will land away from his relatives was likewise recognized."[100]

It should be noted that the term *rist* (*rest*) which in the Abyssinian kingdom denoted family use-rights to communally owned land, was now being used in a different manner. In the southern provinces it referred to private property.

At this time Menilek introduced an Ethiopian currency.[101] Monetization spread slowly and provided an instrument for the wealthy groups to accumulate land.

But during Menilek's time privatization and commercialization of land proceeded slowly. The political decisions marking the definite transfer of *maderia, rist-gult, siso* and other similar tenures to private landownership did not occur until after world war II. They will be treated in chapter 5.

Maintenance of the New Order

During the decades around the turn of the century military violence and the threat of violence was indispensable for the maintenance of the Imperial order in the south of Ethiopia. Soldiers garrisoned in *katamas* constituted the means of coercion. In short, the new order was personified by the Imperial governors who functioned as military commanders, tax collectors and judges. The essential structural aspect of the new order was the *gabar* system and its manifestation through different forms of land tenure.

Menilek was not personally capable of rendering legitimacy to this new order. He remained a stranger and conqueror for his forcefully incorporated subjects. On the local level the order was granted some degree of legitimacy where Menilek utilized *balabbats,* members of the local population as his local officials. But the *balabbats* were not left alone, they were supervised by the Shoan military commander and his soldiers. The function of the *balabbats* was, then, to give legitimacy to the conquerors' orders. The *balabbats* were in relation to the conquerors comparable to "local native chiefs under European administration".[102]

Politically the *gabar* system served Menilek. It enabled him to remunerate his officers and ensure their loyalty at the same time as it provided a system of control (the *katamas*) which kept the young empire tolerably unified. But in economic terms, the *gabar* system was not very profitable for the Crown. Due to the undeveloped infrastructure the local and provincial governors were more or less left on their own. Although information is scanty, it is reasonable to believe that the resources derived from the peasants were mainly consumed by the local and provincial authorities and that only a limited amount of wealth reached the central political authority. Furthermore, the brutal practise of the *gabar* system kept agricultural productivity at a low level.

It is also reasonable to believe that Menilek realized greater economic gain from the areas which had voluntarily submitted to him and thus escaped military occupation. These areas were the Oromo kingdom of Jimma, the Oromo communities in Wollega and the Afar sultanate of Aussa. From these areas Menelik received a yearly tribute in gold, ivory, cattle and agricultural products. Perham states that the arrival of the train of mules and slaves carrying the tribute from Jimma to Shoa was one of the annual events in Addis Ababa.[103]

The chiefs of the peacefully incorporated areas also benefitted. They disappeared as leaders of independent political communities continuously fighting between themselves only to be resurrected in the form of Imperial governors supported by Menilek's supreme authority. At the time of their incorporation, Jimma and the communities in Wollega (especially Quelem and Nakamte) were stratified and the appropriation of land by the ruling groups had progressed considerably. The incorporation strengthened this trend.[104]

The peasants of Jimma and Wollega were also reduced to *gabar* status. They escaped, though, an invasion of Shoan soldiers. It is reasonable to believe that the exploitation of the peasantry in Wollega and Jimma never approached the brutality experienced elsewhere. This was in all probability due to the ethnical, cultural and sometimes genealogical affinity between landlords and peasants.

3.7. Discussion

The main course of events, as they have been described above, will now be discussed in relation to Shoan dominance and methods of surplus appropriation. I will draw on the terminology elaborated by Galtung in his essay on imperialism.[105] Galtung's key terms are *centre, periphery* and *bridgehead.* A social collective (usually a nation) is internally divided into a centre and a periphery (rulers and ruled). One collectivity can exercise relations of dominance upon another collectivity by establishing a bridgehead in the centre of that collectivity. The two centres (the centre in the central collectivity and the centre in the peripheral collectivity) jointly benefit from their dominance over the periphery in the peripheral collectivity. Galtung calls such relationships of dominance imperialism. While utilizing the terms centre, periphery and bridgehead in the ensuing discussion I prefer not to characterize such loosely defined systems of dominance as imperialism. This term gains in scientific clarity if it is used the marxist way: to describe relations of dominance which are established in order to furnish the dominating centre with raw materials and market outlets for its expanding capitalist industries. According to this definition European dominance over African territories can be described as imperialism but Shoan dominance cannot since the appropriation of surpluses from the subdued peripheries to the centre in Addis Ababa occurred largely in the form of paying tribute and were not used for capitalistic accumulation.

The centre in Shoa (i.e. Menilek, his court, the high clergy, military commanders), forcefully established relations of dominance over the earlier independent communities in the south, which then became peripheries of Shoa. The communities on the periphery were penetrated by Shoan bridgeheads, sometimes in the form of military garrisons under a Shoan commander, and sometimes in the form of native rulers who became allied to the Shoan centre.*

Two examples may substantiate the model. i) The kingdom of Jimma was a highly stratified society in itself. Jimma's centre (the king, the landowning *digo* clan and wealthy merchants) appropriated resources from its periphery, the peasants and slaves. The surplus appropriation occurred often under patriarchal forms. When Abba Jifar in 1882 bowed to Menilek, Jimma became a periphery of Shoa. But Shoan troops did not occupy Jimma. Its native rulers served as representatives for Shoan interests. The relations between Jimma and Shoa were organized so that the former paid an annual tribute to the latter in exchange for which its original centre

* It is highly questionable whether the relations between the Shoan centre and the Abyssinian regions in the north fit into this model.

was allowed to remain in power. It can be argued that, under the prevailing circumstances, this arrangement served the interests of both the Shoan and the Jimman centre. The price of the alliance was paid by Jimma's periphery. But it can also be argued that, under the prevailing circumstances, Abba Jifar acted in the best interests of his own periphery. While the 1882 arrangements probably meant heavier exactions from Jimma's peasants, they remained subordinated to the native aristocracy and were spared from brutal oppression by Shoan soldiers. ii) Relations between the Shoan centre and Kafa can be analyzed along the same lines. Kafa was, prior to the conquest, a highly stratified society whose centre (the king, the royal council, the regional governors) appropriated surpluses from the peasants and slaves. After the Shoan conquest in 1897 had eliminated most of the native centre (the king was brought to Addis Ababa in golden chains), a Shoan military commander was installed as Kafa's new centre. The Shoan centre had thereby established a bridgehead through direct dominance. A savage exploitation of Kafa's periphery was soon implemented, the result of which was a decrease in income from tribute for Addis Ababa.

Comparing the two cases, the conclusion can be drawn that it was more profitable for the Shoan centre to use the native centre in the newly acquired peripheries as a bridgehead for the economic interests of the Shoan centre. Existing and well operating mechanisms of surplus appropriation could thereby be geared to enriching the Shoan centre. The use of Shoan military troops as bridgeheads tended to be destructive in the long run because of the crude and savage methods employed by the soldiery. It can thus be argued that there was a conflict of interest present between the Imperial centre and the centres of the periphery in those cases where Shoan military governors and soldiers functioned as bridgeheads. The main objective conflict of interest was, however, undoubtedly between the suppressed peasantry, craftsmen and slaves on one hand and the governors of all levels on the other hand.

Shoan dominance defined in these terms, aimed at enriching its own centre. The mechanism of surplus appropriation was tribute obligations. Within the peripheries, the governors collected tributes, kept part for themselves and sent the remainder to Addis Ababa. Viewed in this way Menilek's empire did not qualitatively differ from the medieval empire whose centre was the court of the Abyssinian king. But since Menilek's empire came into existence in the age of worldwide capitalist expansion it is necessary to ask to what extent the capitalist mode of production was linked to the tributary mode of production.

Menilek's empire was far from isolated from Europe and the Emperor even entertained the desire of 'modernizing' his country. For this reason he nourished ideas about "implanter dans notre pays d'Ethiopie les coutumes européennes".[106] Hess interprets Menilek's intentions thus:

> "Like the rulers of Japan and Thailand, Menilek learned from his experience with Europeans that in order to maintain its independence in the face of European imperialism, Ethiopia must learn much from the imposing body of Western technology. Such knowledge would also be of great usefulness in ruling an empire whose size had doubled in a very short time. For these reasons Menilek employed European advisers to modernize his army, to introduce modern communications and transportation, and to graft onto the traditional semifeudal government more modern bureaucratic institutions."[107]

Pankhurst has given a detailed account of the innovations Menilek initiated after 1896. They included the construction of bridges, the development of a postal system, the building of telegraph and telephone lines, a railway between Addis Ababa and Djibuti in French Somaliland, a few modern roads, a modern hotel, a hospital, a printing press and a secular school in Addis Ababa.[108]

The two authors referred to above mention nothing about 'modernizing' the economy. In fact the main investments were made in physical infrastructure and in urban conveniences for the court and the newly arrived European embassy personnel in Addis Ababa. The agrarian economy remained fundamentally under the tributary mode of production. Only a few rather isolated instances of capitalism occurred in the form of medium scale cotton, rubber, fruit and coffee plantations.[109] These were run by both Ethiopians and foreigners. Menilek also allowed foreign ore prospectors to operate, but their activities did not result in long term, large scale commercial exploitation.[110]

There was no native class of capitalist entrepreneurs in the empire. The leaders of the empire were warlords, landlords and Church dignitaries. A large part of the tribute which accrued to these groups was not convertible into hard currency on the international market, but had to be consumed or used for ceremonial purposes by the receivers of the tribute. The Crown received, however, a number of items (ivory, gold, slaves, beewax, musk) as tribute which were highly valued on the international market. These were thus sold directly as a source of income. The Crown also profited on the private trade in these commodities by levying duties on the individual traders.[111] These duties were payable in Maria Theresia dollars, the internationally accepted currency in northeast Africa, and thus were an important source of international revenue for the Crown.

Estimates of Menilek's revenues and expenditures during the first years of the twentieth century indicate that there were quite large surpluses in gold and Maria Theresia dollars stored in the Imperial treasury.[112] The largest tributes came from Harar, Jimma and Wollega in the south.[113] But the surpluses were reduced by the continuous import of fire-arms, which was an imperative for the survival of the empire and therefore held top priority in the allocation of Imperial resources. Fire-arms could only be paid for in gold and ivory.[114]

Viewed in a continental perspective Menilek accomplished territorial unification and political centralization of the Ethiopian region. This rendered possible a central mobilization of resources utilized to prevent European imperialism from taking over and partitioning the region. Marxist authors such as Endre Sik and Samir Amin have interpreted Menilek's policies in a similar manner, Amin even claiming that Ethiopia embarked on autocentred development.[115] Such an interperation is indeed justified as long as the analysis remains on the continental level. But, when one proceeds to study the relations between the subunits of the Ethiopian empire, one finds that Menilek's policies created uneven development. The relations were structured so that the Shoan centre was enriched and its ruling group became wealthy, while the societies in southern Ethiopia were reduced to dependent peripheries and subjected to an expropriation of resources for the benefit of the Imperial centre. When the analysis is concentrated on the peripheral collectives one can discern internal centres and peripheries, the former acting as bridgeheads for the surplus appropriation of the Imperial centre and the latter (i.e. cultivators, cattle herders and craftsmen) forming the ultimate foundation of the Imperial system.

But during Menilek's time surplus appropriation in the Ethiopian empire did not result in a rapid development of the productive forces. The surpluses exacted were not channeled into investments in commercial production, capitalist relations of production were not introduced on a large scale. In the European colonies in Africa on the other hand, capitalist relations of production were introduced in the mining and agricultural sectors while a growing proportion of the African masses were separated from their means of production. In this important respect the Ethiopian empire was different from the European empires.

Notes to chapter 3

[1] *The Glorious Victories of Amda Seyon,* pp. 36 f.

[2] The districts of Manz, Merhabete, Moret, Doba, Tegulat, Anzokia, Efrata and Geddem are mentioned as the core of eighteenth century Christian Shoa. See Volker Stitz, The Amhara Resettlement of Northern Shoa during the 18th and 19th Centuries, in *Rural Africana,* No. 11, 1970.

[3] Gabra Selase, *Chronique du Regne de Ménélik II Roi des Rois d'Ethiopie* (1930—32), Vol. 1, p. 56.

[4] Levine, *op. cit.*, pp. 31—35.

[5] *Ibid.;* Abir, *op. cit.*, pp. 146 ff.

[6] Gabra Selase, *op. cit.,* Vol. 1, p. 70.

[7] The German missionary Krapf quoted in Abir, *op.cit.*, p. 163.

[8] Abir, *op. cit.*, p. 165.

[9] *Ibid.*, pp. 167, 169.

[10] The British traveller Harris quoted in Richard Pankhurst, *Economic History of Ethiopia 1800—1935* (1968), p. 74.

[11] Abir, *op. cit.*, pp. 162 f.

[12] Pankhurst (1968), *op.cit.*, pp. 579 ff.

[13] *Ibid.*, pp. 2—5.

[14] Abir, *op.cit.*, p. 163.

[15] *Ibid.*, pp. 151—154.

[16] *Ibid.*, p. 51.

[17] Stitz, *op. cit.*, p. 79.

[18] For a biography of Tewodros see Sven Rubenson, *King of Kings, Tewodros of Ethiopia* (1966).

[19] Pankhurst (1968), *op. cit.*, p. 16.

[20] *Ibid.*

[21] For an analysis of the institutions of the pastoral Oromo see Asmarom Legesse, *Gada, Three Approaches to the Study of African Society* (1973).

[22] Huntingford gives an account of the Sidama kingdoms in *Some Records of Ethiopia 1593—1646* and further in G.W.B. Huntingford, *The Galla of Ethiopia; the Kingdoms of Kafa and Janjero* (1955). See also Haberland (1965), *op.cit.*

[23] *Some Records of Ethiopia 1593—1646,* p. lv.

[24] Pankhurst (1968), *op. cit.*, p. 447.

[25] For accounts of the Oromo kingdoms in Gibe see *Some Records of Ethiopia 1593—1646,* Huntingford (1956), *op.cit.*, Trimingham, *op.cit.* and Abir, *op. cit.*

[26] Pankhurst (1968), *op. cit.*, pp. 445 ff.

[27] Trimingham, *op. cit.*, pp. 199—205.

[28] For an account of the administration of Jimma see H.S. Lewis, *A Galla Monarchy: Jimma Abba Jifar, Ethiopia 1830—1932* (1965).

[29] *Ibid.*, p. 50.

[30] *Ibid.*, pp. 50 f.

[31] Trimingham, *op. cit.*, pp. 94, 199; Murdock, *op. cit.*, pp. 324 f.

[32] Negaso Gidada with Donald Crummey, The Introduction and Expansion of Orthodox Christianity in Qelem Awraja Western Wällägä from about 1886 to 1941, in *Journal of Ethiopian Studies* Vol. X, No. 1, 1972, p. 103.

[33] Terrefe Woldetsadik, The Unification of Ethiopia (1880—1935): Wällägä, in *Journal of Ethiopian Studies*, Vol. VI, No. 1, 1968.

[34] Murdock, *op. cit.*, pp. 172 f.

[35] Pankhurst (1968), *op. cit.*, pp. 231 f.

[36] *Ibid.*, pp. 248–252.

[37] Abir, *op. cit.*, p. 86.

[38] Pankhurst (1968), *op. cit.*, pp. 82 ff., 87–92.

[39] Trimingham, *op. cit.*, pp. 203 f.

[40] Pankhurst (1968), *op.cit.*, pp. 199 f.

[41] *Ibid.*, pp. 243 ff.

[42] *Ibid.*, pp. 557–561, 581.

[43] *Some Records of Ethiopia 1593–1646*, p. lxxxvii.

[44] Gabra Selase, *op. cit.*, Vol. 1, pp. 179 f.

[45] Pankhurst (1968), *op. cit.*, pp. 16–19.

[46] Bairu Tafla, Marriage as a Political Device: an Appraisal of a Socio-political Aspect of the Menilek Period 1889–1916, in *Journal of Ethiopian Studies*, Vol. X, No. 1, 1972.

[47] Bairu Tafla, Ras Gobäna Daci, Abba Tegu 1821–1889, in *Journal of Ethiopian Studies*, Vol. V, No. 2, 1967.

[48] Pankhurst (1968), *op.cit.*, pp. 548–552; Habtemariam Seyoum, *Military Organization and Armament Acquisition of Menilek II* (1967), p. 18.

[49] Pankhurst (1968), *op. cit.*, p. 21.

[50] Eike Haberland, *Galla Süd-Äthiopiens* (1963), pp. 530 f.

[51] Tsehai Berhaneselassie, *Menilek II, Conquest and Consolidation of the Southern Provinces* (1969), pp. 4–17.

[52] *Ibid.*, p. 16.

[53] *Ibid.*, p. 17; Gabra Selase, *op. cit.*, Vol. 1, pp. 233–237.

[54] Haberland (1963), *op. cit.*, p. 532.

[55] Gabra Selase, *op. cit.*, Vol. 1, pp. 243–249.

[56] H.S. Lewis, *op. cit.*, pp. 45 f.

[57] For details about the incorporation of the Oromo communities in Wollega with Menilek's empire see Terrefe Woldetsadik, *op. cit.*

[58] Negaso Gidada with Donald Crummey, *op. cit.*, pp. 103 f.

[59] Mahteme Sellassie, The Land System of Ethiopia, in *Ethiopia Observer*, Vol. 1, No. 9, 1957, p. 288.; Terrefe Woldetsadik, *op. cit.*, p. 83.

[60] Margery Perham, *The Government of Ethiopia* (1948), p. 295; Trimingham, *op. cit.*, p. 128; Tsehai Berhaneselassie, *op. cit.*, pp. 24–27.

[61] Gabra Selase, *op. cit.*, Vol. 1, pp. 360–364; Tsehai Berhaneselassie, *op. cit.*, pp. 39–42.

[62] Legesse Gebeyehu, *Conquest of the Kingdom of Kaffa* (1971).

[63] Gabra Selase, *op. cit.*, Vol. 2, pp. 461–464.

[64] Quoted in Amnon Orent, Refocusing on the History of Kafa prior to 1897, A Discussion of Political Processes, in *African Historical Studies*, Vol. III, No. 2, 1970, p. 282.

[65] Legesse Gebeyehu, *op. cit.*, p. 32.

[66] I.M. Lewis, *Peoples of the Horn of Africa, Somali, Afar and Saho* (1955), p. 157. For chronological resumés of the conquests see Perham, *op.cit.*, pp. 294 f. and Trimingham, *op. cit.*, pp. 125–129.

[67] Harold G. Marcus, Imperialism and Expansionism in Ethiopia from 1865 to 1900, in L.H. Gann and Peter Duignan (eds.) *Colonialism in Africa 1870–1960* (1969), Vol. 1, p. 448.

[68] Perham, *op. cit.*, p. 55.

[69] In 1906 an agreement was signed between Great Britain, France and Italy concerning

the partition of Ethiopia into spheres of influence.

[70] The letter is quoted in *The Somali Peninsula, a New Light on Imperial Motives* (1962), Appendix II.

[71] Richard Pankhurst, *The Great Ethiopian Famine of 1888–1892: A New Assessment* (1964).

[72] Gabra Selase, *op. cit.*, Vol. 1, p. 258.

[73] Pankhurst has made an attempt to reconstruct the effects of the famine in detail. See Pankhurst (1964), *op.cit.*

[74] Tsehai Berhaneselassie, *op. cit.*, p. 36.

[75] *Ibid.*, p. 42; Pankhurst (1964), *op. cit.*, pp. 53 f.

[76] For an analysis of the treaty text see Sven Rubenson, *Wichale XVII, the Attempt to Establish a Protectorate over Ethiopia* (1964).

[77] Pankhurst (1968), *op. cit.*, pp. 21 ff.

[78] *Ibid.*, p. 23.

[79] *Ibid.*, pp. 601 f.

[80] Richard Pankhurst, *An Introduction to the History of the Ethiopian Army* (1967), p. 23.

[81] Pankhurst (1966), *op. cit.*, pp. 105 f.

[82] *Ibid.*, pp. 135–148; Gabra Selase, *op. cit.*, Vol. 2, p. 509.

[83] Perham, *op. cit.*, p. 296.

[84] Quoted in Perham, *op.cit.*, p. 320.

[85] Pankhurst (1966), *op. cit.*, pp. 175 f.

[86] *Ibid.;* Mahteme Sellassie, *op. cit.*, pp. 284, 288.

[87] Legesse Gebeyehu, *op. cit.*, p. 44.

[88] Perham, *op. cit.*, p. 319.

[89] Haberland (1963), *op. cit.*, pp. 522, 530.

[90] Perham, *op. cit.*, pp. 319–322.

[91] A.W. Hodson, *Seven Years in Southern Abyssinia* (1927), pp. 242 f.

[92] Pankhurst (1966), *op. cit.*, pp. 107 ff.

[93] MLRA, *Report on Land Tenure Survey of Welega Province* (1968), p. 3.

[94] Mahteme Sellassie, *op. cit.*, p. 285.

[95] Gabra Selase, *op. cit.*, Vol. 2, p. 612.

[96] Mahteme Sellassie, *op. cit.*, p. 285.

[97] *Ibid.*, p. 284.; Pankhurst (1966), *op. cit.*, pp. 136 f., 142.

[98] Pankhurst (1966), *op. cit.*, p. 164.

[99] Gabra Selase, *op. cit.*, Vol. 1, p. 302.

[100] Pankhurst (1966), *op. cit.*, p. 165.

[101] Gabra Selase, *op. cit.*, Vol. 2, p. 456.

[102] Perham, *op. cit.*, p. 307.

[103] *Ibid.*, p. 193.

[104] For accounts of the struggle between local Oromo chiefs and the concentration of power to a few of them through alliances with Menilek and his general Gobana, see Lambert Bartels, Studies of the Galla in Wällägä, in *Journal of Ethiopian Studies*, Vol. VIII, No. 1, 1970.

[105] Johan Galtung, A Structural Theory of Imperialism, in *Journal of Peace Research*, No. 2, 1971.

[106] Gabra Selase, *op. cit.*, Vol. 2, p. 527.

[107] Robert L. Hess, *Ethiopia, the Modernization of Autocracy* (1970), p. 59.

[108] Pankhurst (1968), *op. cit.*, p. 25.

[109] *Ibid.*, pp. 203–208.
[110] *Ibid.*, pp. 231–239.
[111] *Ibid.*, pp. 527 f.
[112] *Ibid.*, pp. 538 ff.
[113] *Ibid.*, p. 537 f.
[114] Pankhurst (1968), *op.cit.* gives many details on Menilek's arms import. Menilek sent special agents to Europe to buy up arms.
[115] Amin (1973), *op.cit.*, pp. 291 f.; Endre Sík, *The History of Black Africa* (1966), Vol. 1, pp. 364, 366.

Chapter 4 Political and Institutional Changes 1913–1941

During the first third of the twentieth century the social structure of the Ethiopian empire did not change qualitatively from the form which it had taken during Menilek's reign. The *gabar* system remained the basis of the empire and surplus appropriation continued to take place under tribute forms.

Quite a number of European and American companies received concessions and were active within the empire. They prospected for minerals and studied the possibilities for agricultural plantations. Very little substance, however, came out of the prospecting activities. Only a handful of large scale, long term investments in production were made.[1] Foreign trade also remained at a low level, imports being dominated by cotton, arms and luxury goods. Export items such as ivory and slaves declined while the export of coffee (mainly from Jimma, Kafa, and Harar) increased.[2] The only innovation with far reaching effects was the Franco-Ethiopian railway from Addis Ababa to Djibuti in French Somaliland.[3] Perham has summarized the achievements on the eve of the Italian invasion.

> "In the last years before the Italian occupation a few innovations had been made by foreigners who introduced a few saw-mills, oil-mills, and soap works. Apart from these beginnings, Ethiopia in 1935 was economically very little changed from what she had been when she was visited by the Portugese in the sixteenth century".
>
> "The old simple system of balanced trade at a low level of government based largely upon revenues in kind and labour, with the large provincial autonomy that this entailed, was a workable system of its own kind".[4]

The political situation was characterized by a halt in the centralizing processes initiated by Menilek. The old Emperor was eliminated from effective government by a stroke in 1907 and died in 1913. From 1907 to 1930 a power struggle between different fractions within the Imperial centre paralyzed effective central government and control over the provinces slackened. As a result provincial governors gained greater independence and central revenues diminished. Gradually Tafari Makonnen, son of Menilek's cousin and loyal servant *Ras* Makonnen, gained the upper hand in the intrigues and struggles. He was appointed regent in 1916 and in 1930 when his most powerful opponents had died or otherwise been eliminated he was strong enough to demand the title of Emperor. He was crowned under the name Haile Selassie.[5]

However progressive the intentions of the young Haile Selassie may have been, he was not in a position to 'modernize' Ethiopia. Intrigues in Addis Ababa and rebellions by provincial governors in the peripheries kept him busy. When he emerged as Emperor in the early thirties, he had to devote all efforts to securing the loyalty of regional authorities and buying up arms under the threat from reactivated European imperialism, this time headed by fascist Italy.

Thus, the central political authority could do little to influence the situation of the peasantry, which was subject to different degrees of oppression by the local and regional authorities. The period is not well documented, and I leave it with these remarks.

Even less space will be devoted to the following period: the Italian occupation 1936—1941. Some main points, however, will be mentioned.

The Italian colonial centre had great ambitions for its newly acquired periphery which, if realized, would have made Ethiopia a settler-colony and created large scale commercial farming in 'white highlands', surrounded by a marginalized Ethiopian population functioning as a labour reserve.[6] In fact, the Italian achievements were of a moderate scale. The most important innovation was a network of roads which for the first time tied the Ethiopian region tolerably together. A number of small scale, light industries were established along the roads.[7]

But the Italians never held effective control over the countryside, and Ethiopian resistance was never completely broken. Guerilla units, 'patriots', operated in various parts of the country, often under the command of members of the previous Imperial centre. But the Emperor himself had chosen to leave the country, a measure which cost him a loss of legitimacy and authority among the Ethiopians.[8]

In 1941 Italian supremacy collapsed. Great Britain chose to break Italian power in the Horn of Africa. British troops entered from Sudan and joined with Ethiopian guerilla forces. In May the Italians had capitulated and Haile Selassie returned from his exile in London.[9]

The complicated power game between the British, Haile Selassie and Ethiopian nobles, which eventually resulted in a restauration of the Ethiopian empire as a politically independent state under the government of Haile Selassie, cannot be analyzed here. Only one point will be elaborated upon. Haile Selassie used land policies as an instrument to gain political supremacy. The analysis in the subsequent chapters will focus on how land manipulations by Haile Selassie's government has brought about structural changes in the southern provinces and how these changes affect the organized living conditions of the peasantry.

Finally a brief account of the institutional framework of Haile Selassie's post-war rule will be given. Under the Emperor there is the Council of Ministers consisting of the Prime Minister and the departmental ministers, all appointed by the Emperor. Since 1966, the ministers are responsible to the Prime Minister and he in his turn to the Emperor. The ministers are heads of their ministries, which numbered 19 in 1966. When the term "the Imperial Government" is used in the following, it refers to the Emperor and the Council of Ministers. No attempt is made to analyse the distribution of power within the Imperial Government.[10]

The regional administration was reorganized. The empire was divided into provinces (governorates general), each province into a number of *awradjas* and each *awradja* into a number of *woredas*. Each administrative unit is headed by a governor appointed by the Emperor and paid a fixed salary from the central treasury. The governors are assisted by vice-governors and secretaries and they command a police force. The line of command is hierarchical, the *woreda* governors report to the *awradja* governors who report to the provincial governors. The formal authority of the provincial governors is the Ministry of Interior. Below the level of the *woreda* the semi-official government continues, the hereditary office of *balabbat* being subordinate to the *woreda* governor.

Furthermore there is a bi-cameral Parliament which, together with the Emperor performs the law-making functions.[11] The members of the upper house, the Senate, are appointed by the Emperor, while the members of the lower house, the Chamber of Deputies, are elected. The formation of political parties for elections is forbidden. Parliament is filled with landlords of various magnitudes. It by and large sanctions decisions already taken by the Imperial Government. In cases when Parliament has articulated an opinion of its own, it has functioned as an agent for the interests of the landlords.[12]

Notes to chapter 4

[1] Pankhurst (1968), *op. cit.*, pp. 208 f., 231–243.

[2] *Ibid.*, chapter "Trade".

[3] Perham, *op. cit.*, p. 194.

[4] *Ibid.*, pp. 180, 199.

[5] Christopher Clapham, *Haile-Selassie's Government* (1969), pp. 15 f.

[6] Richard Pankhurst, A Page of Ethiopian History: Italian Settlement Plans during the Fascist Occupation of 1936–41, in *Ethiopia Observer*, Vol. XIII, No. 2, 1970.

[7] Perham, *op. cit.*, pp. 180–185.

[8] For an account of the Ethiopian 'patriot' forces and their relations to Haile Selassie see Richard Greenfield, *Ethiopia, A New Political History* (1965).

[9] Clapham, *op. cit.*, p. 19.

[10] For a detailed analysis of the Imperial Government see for example Clapham, *op. cit.*

[11] For a formal review of the functions of the Parliament see Kenneth Redden, Aberra Jembere and Nabiyelul Kifle, *The Law Making Process in Ethiopia* (1966).

[12] For a case study of the 1967 legislation on agricultural income taxation, when Parliament successfully challenged the Imperial Government's proposition see Peter Schwab, *Decision-Making in Ethiopia, A Study of the Political Process* (1972).

Chapter 5 The Imperial Government's Land and Agricultural Policies after 1941

5.1. Imperial Land Grants

The Imperial Orders

After the restauration of Imperial rule in 1941 Haile Selassie activated the age-old Ethiopian policy of granting land to politically important individuals and groups. Haile Selassie rendered new features to the grants. Earlier mainly temporary rights had been granted to exact tribute from the produce of the peasants, but now the Emperor started giving freehold status to the grants. This was accomplished through a series of Imperial Orders concerning unutilized Government land as well as Government land which already was held under *maderia* and similar forms of tenure. The most important Orders are summarized below.[1]

1942

Patriots and exiles who had served loyally throughout the occupation were entitled to a freehold grant of one *gasha**. Those who already held Government land under *maderia* were entitled to convert one *gasha* thereof to freehold.

1944

Soldiers and civil servants who had served the Government up to the invasion and who did not hold *maderia* land were entitled to a freehold grant of one-half *gasha*.

1952

Landless and unemployed Ethiopians were entitled to a freehold grant of one-half *gasha*. All holders of *maderia* land were entitled to convert one-third *gasha* to freehold (in addition to the conversion right given in 1942).

1956

All *maderia* holders were entitled to convert an additional two-third *gasha* of their *maderia* land to freehold.

* One *gasha* = 40 hectares.

1959

All members of the army, air force, the Emperor's body guard and police were entitled to freehold grants of one *gasha*. In 1965 this right was extended to all civil servants as well.

1966

Special rules were provided for further conversion of *maderia* land to freehold in Wollo province.

1967

Maderia holders within the territorial army were entitled to convert additional amounts of their *maderia* land to freehold according to the following; captains: 4 *gashas*, lieutenants: 3 *gashas*, sergeants: 2 *gashas*, all other ranks: 1 *gasha*.

Application of the Imperial Orders

The grants were not given automatically to those theoretically entitled to them, but were made dependent upon a written application from the individual grantees.[2]

Up to 1969 about 44,000 *gashas* (1, 760,000 hectares) of Government land had been granted to individuals in accordance with the various Orders, and that same year there were applications pending for land amounting to a total of 62,000 *gashas*.[3] From 1969 onwards the granting processes have continued. The land granted is mainly in areas of the southern provinces, where the bulk of the Government land is situated.

It is above all Government employees (members of the military and civil servants) who have benefitted from the Imperial Orders.

About 80 % of the land grants given up to 1969 had gone to members of the armed forces and to civil servants.[4] Landless and unemployed persons had received only a "negligible amount".[5] A case study of the social background of those who had received grants of two *gashas* or more showed that 80 % of the grantees held titles as *dedjazmatch, leul ras, fitawrari, bejrond, kantiba, afanegus* etc.,[6] which means that they belonged to already privileged groups.

The application procedure for grants treats poor people disfavouringly. When it comes to the Order regulating grants to landless and unemployed, it is the applicant himself who must prove to the *woreda* officials that he is unemployed and landless.[7] If the applicant is successful in convincing the local officials about his status, these issue a certificate that he is qualified to apply for land according to the 1952 Order.[8] Thereafter correspondence with upper strata of the bureaucracy commences. The outcome is, however, uncertain.

> "It has been necessary for such a person to work through a long chain of officialdom to reach his target . . . the unsponsored landless and unemployed applicant has had a long and expensive struggle with bureaucracy. Many applicants with insufficient stamina to stay the course have dropped out and many more, deterred by the experience of their fellows, have not even started".[9]

The quotation hints at the corruption which prevails on the lower levels within the administration. Unemployed and landless persons simply cannot afford all the bribes which are necessary in order to make the *woreda* governor and secretaries co-operative.

Members of the military, police and civil servants do not encounter the same hindrances. Their application procedure is made easier by a certifacte of their qualifications issued by their Minister.[10] The applications are channeled to the Ministry of Land Reform and Administration, which since 1968 administers applications for land grants. After transmission to the local authority in the *woreda* where the land applied for is situated, the Ministry authorizes the grant.[11] But it would be erroneous to believe that the granting process is smooth even for these more privileged groups of applicants.

> "Once a certificate for grant is ascertained, an endless correspondence starts between the provincial office and the lower administrative levels, as a result of which much time and effort is sacrificed".[12]

In some cases a grantee's application had to pass 40 officials before it was finally approved.[13] The process is time consuming. According to an investigation some applicants have waited from four to eight years to secure a grant. The delay is caused above all in the *woreda* and *awradja* offices where applicants must spend time and money on bribing officials to write certificates and bribing *balabbats* to point out where the Government land is situated.[14]

All this indicates the limited control exercised by the Government at the lower administrative levels. The local governors have a considerable margin within which they can promote thier own interests, despite the formally high degree of political centralization.

Utilization of the Grants

Some of the Orders were accompanied by the Imperial appeal; "Develop the land on receiving it from Our delegates".[15] But since no concrete programme for agricultural development followed, the appeal remained empty of content.

Since the bulk of the grantees are urban Government employees, few of them have settled on their new holdings. A sample survey on how the granted land in five provinces had come to use, showed that at the time of the survey only 1/3 of the granted land included in the survey was under cultivation, and 2/3 of this area was not cultivated by the grantees themselves but by tenants.[16] The general pattern is summarized thus:

> "It is widely acknowledged that a very small proportion of the recipients of patronage grants actually farm the land themselves. This may be because the grants are generally in areas remote from their homeland and also remote from the area in which they are serving. Sometimes grant land is sold to local people but generally the recipient retains it himself as a landlord and takes in tenants, farms it on a share cropping basis or lets it to graziers".[17]

Formally the grants can only be made from unutilized Government lands. In reality there are few areas in the country with an agricultural potential which lie unutilized. In Wollega the Ministry of Land Reform and Administration found that on lands formally classified as "Government unoccupied land" there were a great number of peasants "illegally cultivating" the land. In 1971 alone, 15,700 "illegal cultivators" were registered on Government lands in Wollega.[18] When grantees inspect their newly acquired freehold land and find peasants living there, they establish themselves as landlords over these people and start demanding rent.[19]

The Emperor's Special Grants

The Imperial Orders commented on above concern broad categories of individuals. But there is also a procedure through which land can be granted to private persons. The Emperor can grant Government land to persons "whose services to the empire have earned them His Majesty's special grace".[20] This happens through the Special Grant Order.[21] The amount of Government land distributed in accordance with this Order is not publicly known.[22] But since this Order provides an instrument for the Emperor to secure the loyalty of persons whom he considers politically important, it is reasonable to believe that Special Grants have been made for large areas of fertile land.

The Land Grants as Political Instruments

The officially stated motives for the Land Grant Orders are based on care of the material welfare of the grantees. The Emperor motivated the 1959 Order thus:

"The purpose in granting you land, besides increment in pay, is to enable you to have a lasting capital which can also be inherited by your children".[23]

It can be argued that there are political considerations intended to strengthen Imperial power lying behind this facade.[24] The Orders of 1942 and 1944 were issued shortly after the defeat of the Italians. During these years Great Britain was militarily present in Ethiopia and Haile Selassie's position as supreme political authority was far from undisputed. British troops remained in the country and among the Ethiopian guerilla fighters there were groups who questioned Haile Selassie's role during the occupation and considered him a traitor.[25] Consequently, the two first Orders can be interpreted as attempts to win the loyalty of those who actively had resisted the invaders and who therefore commanded influence and popularity. The Orders of later dates have continued to favour the military and the civil service for the same political reasons; to strengthen the loyalty of these important groups to the grantor and to prevent smouldering discontent from bursting into flames.

The motive behind the 1952 Order might well have been to emphasize Imperial concern for the poor and landless. The fact remains, however, that only a fractional part of these Ethiopians, who constitute the majority of the population, have been able to benefit from the promises of the Order.

Structural Implications of the Imperial Land Grants

The importance of the Imperial Orders regulating land grants was that they initiated private land ownership on a large scale in southern Ethiopia. The Government used its theoretical claim of overlordship over all land as an instrument for transferring wast amounts of land from peasant and pastoral communities into private hands. The individuals who received land belonged primarily to groups which the Emperor had to remunerate, lest they should challenge his political supremay.

The orders worked in two directions. On one hand they created peasant proprietors. Landless Ethiopians who managed to receive ownership rights through the 1952 Order and soldiers who managed to get pieces of land according to the various Orders took up agriculture themselves or partitioned the land among their children. On the other hand the Orders created a group of large, mostly absentee, landowners. Those who received grants amounting to several *gashas* of land, in particular the beneficiaries of the Emperor's Special Grants, were people who already held important positions. They did not take up agriculture themselves, but demanded rent from the peasants and pastoralists living on their private property.

An attempt to estimate the number of owner-cultivators and absentee landowners respectively would prove to be arbitrary. Therefore, suffice it to say that although the peasant proprietors certainly form a quantitatively much larger group than the absentee owners, the structural trend in the 1950's, 1960's and early 1970's has been in the direction of concentrating land ownership in fewer hands.

The consequences of the land grants for the peasants were that while a minority of peasants became owner-cultivators, the majority was transferred from a status of *gabars* to tenants. The *gabar* system was officially abolished in 1941.[26] The juridical change from ownership of land by the Imperial Crown to private ownership meant little in the daily life of these peasants. They were still the subjects of military officers, civil servants, *balabbats* etc. who now emerged as the sole owners of the land which they cultivated as tenants.

In southern Ethiopia landlords and tenants constitute the two principal social classes in relation to the ownership and control of land, which is the main means of production. In between them there are the owner-cultivators. Those owner-cultivators who own only a few hectares and perhaps rent additional land have much in common with the tenants. Together they constitute the poor peasantry, which is the overwhelming majority of the agricultural population. There is also a group of owner-cultivators who own more land than they can cultivate themselves (using family labour and work oxen). They lease out their additional land to tenants and receive incomes from land rent. The upper stratum of this group passes smoothly into the lower stratum of the landlords, i.e. persons who do not practise agriculture themselves, but live from land rents.*

Expressed in the analytical language of modes of production, the land grant Orders signified an important vehicle in the process whereby the tributary mode developed into feudalism. Consequently, I will refer to the social structure in the southern provinces of Ethiopia after the 1940's as predominantly feudal.

5.2. Legislation on Land Taxation

The changes in the ownership pattern of land were accompanied by new legislation on land taxation. Up to 1942 taxes had been paid in kind and had been levied arbitrarily. The land tax legislation of 1942 and 1944 laid down a more precise pattern for land taxation, which was intended to

* For further classification of peasants and landlords see chapter 6.

regulate and increase Government revenues. This legislation has been described and discussed in detail by Peter Schwab.[27] Here, only the main points of the legislation will be reviewed.

It was stated that taxes should be levied in money per *gasha* per annum. In areas where the land was unmeasured the rate was fixed to Eth. $ 12 per *gasha*, in measured areas* it varied from Eth. $ 5 to Eth. $ 15 according to the estimated fertility of the soil. In addition to the land tax, tithe should be paid in money. The tithe was fixed to Eth. $ 8 per gasha in unmeasured areas and to a sum varying from Eth. $ 10 to Eth. $ 35 in measured areas. It was the intention of the tax proclamations that the taxes should be paid by the landowners.[28]

Initially the legislation was a success for the Government. In 1944/45 the land tax earned Eth. $ 3.9 million and the tithe Eth. $ 5.7 million for the State treasury. Together these sums amounted to about 20 % of total Government revenues.[29] But in the longer perspective the legislation proved to a failure. The revenues from land tax and tithe have increased only marginally. In 1961/62 the land tax amounted to Eth. $ 5.2 million and the tithe to Eth. $ 10.3 million, while in 1966/67 the corresponding figures were Eth. $ 6.3 million and Eth. $ 10.6 million. During the same period Government expenditures have increased tenfold,[30] which means that revenues from land tax and tithe have declined considerably in importance for Government finances.

Among the main reasons for this failure were: i) the new system of taxation was limited to the southern provinces. In Tigre, Beghemder and Godjam provinces the old system of levying a fixed sum on a province as a whole was retained, ii) land belonging to the Church was exempted from taxes to the Government. iii) most lands remained unmeasured and were thus subject to the softer tax rates. The Government did not have precise information on the size of the larger holdings which meant that owners could report less land than they actually owned.[31]

But the landowners still considered the tax rates to be too high. Under pressure from the landowners the Government issued amendments to the legislation in 1951 and 1962. The amendment of 1951 stipulated different tax rates according to estimated fertility also on unmeasured lands. The lowest land tax rate was fixed to Eth. $ 2 per *gasha*, per annum and the lowest rate for tithe to Eth. $ 3.[32] No guidelines were provided for estimating the fertility of the soil. The tax assessment procedure was also changed. According to the Proclamation of 1944 the local tax assessment

* Up to 1966 only about 10 % of the area which has an agricultural potential had been measured.

committees were to be dominated by representatives of the Ministry of Finance. According to the Legal notice of 1951 the assessment committees were to have 7 members, of which only 2 members represented the ministries, the others being prominent local people.[33] Thus, the field was open for systematic tax evasion.

Furthermore, the rule has been that the landowners pay only the land tax, and make their tenants pay the tithe. This had been praxis since the time of Menilek and the 1944 stipulation that the landowners should pay the tithe has not been communicated to the tenants. Actually, tenants and owner-cultivators have contributed the bulk of Government revenues accrued from land tax and tithe. Concerning the burden of taxation a FAO report states that "the custom would appear to penalize farmers, both owner farmers and tenant farmers, and to favour absentee landlords".[34]

Since the Government provided the large landowners with loopholes for tax evasion in the application of the land tax legislation and since it did not interfer to prevent the landowners from shifting the main burden of taxation (tithe) to their tenants, it can be concluded that the Government systematically favoured the landlord class, even at the expense of its own revenues. Land taxation did not primarily function as a mechanism of accumulation of resources for the Imperial centre, because that would have clashed with an objective of higher priority for Haile Selassie's Government—that of preserving the political loyalty of the large landowners.

In 1967 the process was repeated. The Ministry of Finance embarked on a renewed offensive in the field of land taxation. Also this time the original aim was to increase the taxation of large owners. But the Government eventually retreated when confronted by fierce opposition from the large landowners, who this time used parliamentary action to thwart the legislation. The result—the Agricultural Income Tax law of 1967—emphasized taxation of tenants and owner-cultivators.[35] Consequently, Government revenues from land taxation has remained on a low level.

5.3. Development Policies

The Political Framework

In the early 1940's the Imperial Government expected much from land tax legislation. The land taxes would raise Government revenues considerably it was argued.[36] But these expectations proved to be false. Still, if Haile Selassie was to be able to regain political supremacy and carry on the processes of centralization, it was imperative for him to command resources of a magnitude far beyond pre-war governmental revenues. Resources on such a scale could not be mobilized from the agrarian economy of Ethiopia solely through refining the mechanisms of surplus appropriation. Particularly when the Government chose not to challenge the regional and local power-holders' control over the peasantry. The tributary mode of production, which in earlier historical epochs had nourished complex political structures, was in the post-world war period obsolete and insufficient.

Haile Selassie turned to foreign sources for capital. In 1950 the Imperial Government received its first loan from the World Bank group* and in 1953 from the US Government.[37] These events were indicative of coming developments. The US Government and the World Bank group have become the main sources of capital for the Imperial Government.

Two trends can be observed in the pattern of foreign assistance to the Imperial Government. One is the provision of immediate resource stabilizers, such as a mechanized warfare apparatus and financial credit, which have been indispensable for the short term operation of the Imperial Government; rebellions have been crushed, imports of industrial and luxury goods have been increased. The other trend is assistance for the long term objective of initiating a capitalist mode of production within Ethiopia aimed at accelerating the accumulation of domestic resources. This kind of assistance has taken three forms: i) investments in the infrastructure aimed at opening up the countryside, ii) direct investments in productive installations by foreign capital, sometimes in partnership with the Imperial Government and under the auspices of the World Bank group, iii) financial and technical assistance to promote indigenous Ethiopian enterprises.

The capitalist road to development embarked upon by the Imperial Government has been codified in five year development plans which express the official Ethiopian development strategy. The Third Five Year Development Plan 1968–1973 (TFYDP) stresses "the fastest possible growth and development of the economy as a whole", and proposes capital investments amounting to Eth. $ 3,115 million.[38] This sum of money cannot be raised within the country. TFYDP emphasized that

> "If the development plan is to be successfully implemented there is no escape from the necessity for Ethiopia to take on the large additional amounts of external capital obligations. The Government is therefore deeply concerned about obtaining the required loans,

* The World Bank group includes the International Bank for Reconstruction and Development (I.B.R.D.) and the International Development Association (I.D.A.).

credits and investments—on terms as favourable as possible. It will therefore seek from external aid givers, particularly official multilateral and bi-lateral sources, the maximum grace and repayment periods possible for loans and credits, as well as the most favourable interest rate terms".[39]

The Emperor has not neglected to communicate the same message:

> "We have never ceased to encourage foreign capitalists to work either in co-operation with Our people, or to invest their money on development works on their own. It has for long been part of the tradition of Ethiopia to welcome foreigners and to safeguard their rights and interests during their stay here. In the case especially of those foreigners who have invested capital on development projects in this country, apart from working in co-operation with Our Government officials and Our people, there has never been a case in Ethiopia's history when the property of a foreigner has been unlawfully confiscated. This is a source of pride for Ethiopia. Ethiopia will at all times welcome foreign investors who may be willing to employ their money for development works, either in partnership with Ethiopia, or as private enterprisers".[40]

Legislation on foreign capital investment was issued in 1966. The Investment Proclamation guarantees exemption from payment of taxes during the first five years for foreign companies which invest more than Eth. $ 200,000 in Ethiopia. The Proclamation stipulates that the National Bank of Ethiopia shall see to the remittance of profits abroad by changing them into hard currency. If the foreign company decides to conclude its business activities, all of its assets in Ethiopia may be converted into hard currency.[41]

It can be argued that Ethiopia, starting in the post world war II period, is in a process of being drawn into the international capitalist system and reduced to a dependent periphery of the centre (the United States, Western Europe, Japan) in this system. The Imperial Government has functioned as a bridgehead for the interests of the centre in the international capitalist system. It has opened up the country for its external allies. An American military base (specialized in international telecommunication) was established in Asmara and it has functioned in the Vietnam war. US, Western European, Japanese and Indian business interests have invested in plantations, oil, gas and mineral prospection and, to a less extent, in the manufacturing industry.

There seems to exist a harmony of interests between the foreign companies and the US Government on one hand and the Imperial Government on the other. The US Government has acquired a strategical military foothold on the roof of Africa and foreign companies have gained access to cheap labour and natural resources in a good 'investment climate'. The Imperial Government has received in return a technically advanced military apparatus to suppress internal opponents and in addition the members of the Imperial family as well as members of the Government have acquired shares in the companies which have established themselves in the country.

By functioning as a bridgehead for the international capitalist centre, the Imperial Government can compensate for revenues both unattainable and forgone at home. The low amount of domestic revenues accrued to the Government can, in part, be explained by referring to the Government's internal alliances. At home the Government is dependent on support from the Ethiopian Orthodox Church* and from the landlords. To maintain their loyalty the Government has chosen not to increase the taxation of their incomes. In spite of the avowed high degree of centralization in the country, the landlords (in the southern provinces) and the Church (in the northern provinces) are the powerholders on the local level.

In the following discussion the framework outlined here should be kept in mind. The discussion will, however, be successively narrowed to include only the relations between the Imperial Government, the landlords and the peasants in the cereal producing highlands of Ethiopia's southern provinces.

The Importance of Agriculture

While in the 1950's and the first half of the 1960's priority was given to improving the infrastructure, in the second half of the 1960's the strategical role of agriculture started to be emphasized. A World Bank report from 1967 pointed out that

> "The urgent task facing Ethiopia is the initiation of measures to modernize its agriculture. Without a major effort to increase investment in agriculture substantially, to train more people, apply research and know-how and improve organization for agricultural development, a major expansion in Ethiopia's economy cannot be expected".[42]

The Prime Minister of the Imperial Government supported this concept in the foreword of TFYDP:

* On the national level the Ethiopian Orthodox Church is politically important. On the local level the importance of the Church is also great in the northern provinces where the bulk of its landed interests are located. In the southern provinces the Church owns comparatively little land and this holds true also for the areas which will be studied in detail in the following chapters. Therefore, the role of the Church is left out of this analysis.

"The First Five Year Plan had given priority to infrastructure. The Second Five Year Plan gave greater emphasis to directly productive undertakings, particularly manufacturing industry, mining, electricity, etc. In the Third Five Year Plan there is a substantial shift of emphasis from the preceeding plans by giving high priority to agriculture".[43]

The agricultural sector is considered important for the implementation of TFYDP.

"Attainment of the national goals necessarily involves highly variable targets of expansion among the different sectors of economic activity. The foundation for the Ethiopian economy is and must long remain agriculture. But the development of agriculture and especially of agricultural society in the proper broad context of this plan, requires emphasis also on a concurrent development of industrialization. In the Ethiopian context this means especially the development of those lines of industry most closely related to agriculture and normally included under the heading of agroindustry. The real and sustained development of Ethiopia is, in any event, unthinkable without sound progress in the expansion of agricultural output".[44]

An increasing agricultural production is needed in order to

"... produce steadily expanding supplies of food-stuffs to meet the growing money demand for food from its rapidly increasing urban population. The present nutritional levels of much of the population must be improved. Ethiopia needs also to expand agricultural exports in order to pay for the growing volume of imports demanded by a growing economy".[45]

In TFYDP, agriculture is divided into 'commercial agriculture' and 'peasant agriculture'. The former includes sugar-, cotton-, tobacco-, oilseed- and coffee-plantations, preferably in the lowlands, and large scale cereal producing farms in the highlands. As the name indicates commercial agriculture is run as business enterprises producing for the market.

Peasant agriculture comprises the low-yielding, family-based, small scale, predominantly subsistence (and tribute) oriented agriculture performed by the approximately 4 million peasant households in the highlands of the empire.

Development Strategy for Large Scale Commercial Agriculture

A point of departure for the planners of TFYDP was the impossibility of generating radical changes within the peasant agricultural sector during the five year period (1968—73) and therefore their interest was focused on promotion of commercial agriculture. The reasoning took the following form:

> "The rapid development of commercial agriculture is the only way to get the relatively quick increase needed in agricultural exports. It will clearly be essential to induce more foreign private investment and to import the needed managerial and technical skills; these farming enterprises may be public or private in ownership and operation but the really important consideration is that the activities be commercially and financially sound. It is from this sector that the rapid gains are expected in output and availability of surpluses, both for consumption domestically, particularly in the cities and towns as well as for export."[46]

Commercial production of cotton and sugar in particular needs heavy investments and therefore foreign private capital was encouraged in this field, alone or in co-operation with the Imperial Government. Commercial production of cereals in the highlands is less capital demanding, therefore private Ethiopian citizens have had an opportunity to establish themselves as entrepreneurs.

The analysis which follows in the succeeding chapters is limited to the southern highlands. Therefore the commercial agriculture established in the Awash Valley (mainly Dutch, British and Italian capital) and in Setit-Humera (private Ethiopian entrepreneurs with World Bank support), which represent huge economic interests and has caused dramatic contradictions, will not be analyzed here. Instead the analysis will focus on the contradictions between mechanized commercial farming and peasant agriculture in the southern highlands.

It remains now to enumerate the different subventions given to private persons who intend to establish commercial farms. These subventions include:[47]

i) credits in kind for fertilizer

ii) credits in cash (interest rate: 9 %) for purchase of agricultural machinery (tractors, trailers, combiners)

iii) purchase of duty-free fuel and machinery for agricultural purposes (duty-free fuel ended in June 1973)

The institution which provides these subsidies is the Ethiopian Agricultural and Industrial Development Bank (AID Bank). The AID

Bank demands security in land or cash for the credits,[48] which has tended to restrict its clientel to already wealthy persons.*

Development Strategy for Peasant Agriculture

TFYDP bluntly states that "there is no quick solution to the peasant problem".[49] Therefore the policies concerned with the peasants have the character of long-term projects.

> "Modernization of peasant subsistence agriculture in all areas of the country simultaneously is hardly feasible. It would merely mean dilution of effort and of limited resources. But no time should be lost in making a start in strategically selected areas in which good results can soon be seen".
>
> "The most promising approach to the successful development of strategically selected areas is that known as the 'package programme'. This simply means the policy of concentrating development efforts, in given areas of not too great a size, utilizing all the material and human resources required to change traditional and inefficient methods. Part of the objective is to increase output so much that significant surpluses become available for marketing outside the area; as a result, the peasants will be introduced to the commercial market system."
>
> "To complete the package programme, the farmers must be assured of a supply of incentive consumer goods on which they can spend newly acquired, or increased, money incomes. The number, kind and price of these must be such as to maintain their interest, so that there is an incentive to earn even more".[50]

The essence of this strategy is to force the Ethiopian peasantry into the capitalist mode of production. It is assumed that through the 'package' approach, this will improve the living conditions of the peasantry as well as create a solid basis for Government tax revenues.

The package approach to developing peasant agriculture will be analyzed in chapters 7 to 10. Here it is sufficient to say that in practice this approach has concentrated on providing peasants with fertilizer, in order to increase their cereal production, and on promoting the establishment of co-operative societies for the marketing of the increased cereal production.

Structural Contradiction between the Strategies

If large scale commercial farming was only promoted on hitherto un-

* The AID Bank is also involved in the provision of credits to peasants through the CADU project and the minimum package projects. See chapter 7.

utilized lands and peasant agriculture was promoted in the traditionally cultivated areas, then there would exist no immediate contradiction between the simultaneous promotion of the two strategies. But this has not been the case in Ethiopia. In fact, large scale commercial farming and peasant agriculture have been promoted in the same geographical areas, i.e. areas where the land for a long time has been under peasant cultivation. Since the later part of the 1960's this process has been underway in a growing number of areas in the southern highlands. In these areas a contradiction has developed between the promotion of large scale commercial farming and peasant farming. The land is already being utlized for crop growing or for grazing by the peasants. If large scale commercial farms are to be established, the peasants must yield. If peasant agriculture is to be promoted, there is no room for large scale farms. In these areas both commercial farmers and peasants cannot flourish side by side, one strategy supporting the peasants and the other the commercial entrepreneurs. Since both strategies are dependent on governmental subsidies, the actual outcome of the confrontation can be taken as an indicator of which party the Imperial Government favours.

Notes to chapter 5

[1] Summaries of the Orders can be found in MLRA, *A Policy Oriented Study of Land Settlement* (1969), Annex IV; Negatu Lakew, State and Land Grant System in Ethiopia, in *Seminar Proceedings on Agrarian Reform* (1970); Yoseph Haile, *Relationship between Government Land and Political Power in Ethiopia* (1973).

[2] Yoseph Haile, *op. cit.*, p. 20.

[3] Hans Wetterhall, *Memo on Government Land in Ethiopia* (1968).

[4] Harrison Dunning, Land Reform in Ethiopia; A Case Study in Non-Development, in *UCLA Law Review*, Vol. 18, No. 2, 1970, p. 290.

[5] *A Policy Oriented Study . . .*, Vol. I, p. 21.

[6] Belayneh Negatu, *The Use of Development Conditions on Government Land Grants in Ethiopia: Practise and Prospects* (1969), pp. 23 f.

[7] Negatu Lakew, *op. cit.*, p. 213.

[8] *Ibid.*

[9] *A Policy Oriented Study . . .*, Vol. I, p. 22.

[10] *Ibid.;* Negatu Lakew, *op. cit.*, p. 213.

[11] Belayneh Negatu, *op. cit.*, p. 17.

[12] Negatu Lakew, *op. cit.*, p. 214.

[13] Information obtained from the Ministry of Land Reform and Administration.

[14] Belayneh Negatu, *op. cit.*, p. 35.

[15] *Ibid.* Appendix I and II.

[16] MLRA, *Policy of the Imperial Ethiopian Government on Agricultural Land Tenure* (1972).

[17] *A Policy Oriented Study . . .*, Vol. I, pp. 21 f.; Cf. Aberra Moltot, The Need for Public Participation in the Executions of Land Reform Laws, in *Seminar Proceedings on Agrarian Reform* and Negatu Lakew, *op. cit.*

[18] *Policy of the Imperial Ethiopian Government on Agricultural Land Tenure.*

[19] *Ibid.*

[20] *A Policy Oriented Study . . .*, Vol. I, p. 21.

[21] *Ibid.*; Belayneh Negatu, *op. cit.*, p. 18; Yoseph Haile, *op. cit.*, p. 17.

[22] Negatu Lakew, *op.cit.;* A sample investigation showed that approximately 3/4 of the beneficiaries of the Emperor's Special Grant Order were 'big men'. Yoseph Haile, *op. cit.*, p. 22.

[23] Belayneh Negatu, *op. cit.*, Appendix I.

[24] Such an interpretation is also the main theme in Yoseph Haile's and Belayneh Negatu's essays.

[25] Greenfield, *op. cit.*

[26] Perham, *op. cit.*, p. 355.

[27] Schwab, *op. cit.*

[28] Ibid., pp. 27 f., 40 f., 45; J.C.D. Lawrence—H.S. Mann, F.A.O. Land Policy Project (Ethiopia), in *Ethiopia Observer,* Vol. IX, No. 4, 1966, p. 320.

[29] Perham, *op. cit.*, pp. 202 f.

[30] *Ibid.;* CSO, *Statistical Abstract 1963,* pp. 90 f.; CSO, *Statistical Abstract 1970,* p. 130.

[31] Schwab, *op. cit.*, pp. 31, 40 ff., 47 f.; Lawrence—Mann, *op. cit.*, pp. 320—324.

[32] Negarit Gazeta, *Proclamation No. 117 of 1951. A Proclamation to Amend the Land Tax Proclamation of 1944.*

[33] Negarit Gazeta, *Legal Notice No. 154 of 1951.*

[34] Lawrence—Mann, *op. cit.*, p. 322.

[35] Negarit Gazeta, *Proclamation No. 255 of 1967. A Proclamation to Amend the Income Tax Proclamation, 1961;* For a detailed analysis of the political intrigues which gave the law its definite formulation see Schwab, *op. cit.*

[36] Perham, *op. cit.*, p. 207.

[37] IBRD, *The World Bank Group in Ethiopia* (1973), p. 3.

[38] *Third Five Year Development Plan (1968—1973)* (1968), p. 59 (Hereafter: *TFYDP*).

[39] *Ibid.*, p. 78.

[40] Haile Selassie I, *Selected Speeches* (1967), pp. 502 f.

[41] Negarit Gazeta, *Proclamation No. 242 of 1966. A Proclamation to Provide for the Encouragement of Capital Investment in Ethiopia.* The proclamation is based on *Decree No. 51 of 1963. A Decree to Provide for the Encouragement of Capital Investment in Ethiopia.*

[42] IBRD, *The World Bank Group in Ethiopia* (1967), p. 3.

[43] *TFYDP,* p. 19.

[44] *Ibid.*, pp. 36 f.

[45] *Ibid.*, p. 181.

[46] *Ibid.*, p. 191.

[47] Information obtained from the Public Relations Office of the AID Bank.

[48] *Ibid.*

[49] *TFYDP,* p. 190.

[50] *Ibid.*, pp. 193 f.

Chapter 6 An Appraisal of the Land Tenure Situation in the 1960's

6.1. The Land Use Pattern

The total area of Ethiopia is 122.2 million hectares. Estimates for 1970/71 classify some 11 % of the area (13.5 million hectares) as cultivated land, 54 % as pastures, 10 % as water courses, 7 % as forests, 4 % as swamps and 14 % as barren land.[1]

The bulk of the cultivated land as well as the peasant population is found in the rainfed highlands. Most crops grown here are cereals (barley, teff, maize, sorghum and wheat), more than 70 % of the cultivated area being devoted to cereals. Another major crop, ensete, is largely confined to the Sidama and Gurage regions in the southwestern highlands.[2] Coffee, the most important export crop, is grown in selected parts of Kafa, Sidamo, Wollega, Illubabor and Hararge provinces. Coffee is grown on less than 10 % of the total cultivated area.[3]

Fallow land was estimated to comprise more than 2 million hectares in the late 1960's, which means that only somewhat more than 10 million hectares were actually planted and harvested.[4]

The farming technology of the cereal cultivating peasants is a product of innovations brought to the Ethiopian highlands some time before recorded history. Over the centuries, agricultural techniques have remained on more or less the same level. The ploughing of fields is performed with an ox-drawn wooden plough with metal edge. Hoes are also used, in some parts of the south hoe cultivation being the rule. Irrigation and terracing is practised in many places. Reaping of crops is done by sickle, threshing by driving oxen over the sheaves, and winnowing by throwing the threshed grain into the air so that the husk is removed by the wind.[5]

Under favourable conditions a peasant family can cultivate a maximum of 10 hectares using these methods and tools.[6] Average yields per hectare are then: teff 5 to 7 quintals, wheat 6 to 8 quintals and maize 9 to 11 quintals.[7]

Crop cultivation is supplemented by animal husbandry. The peasants keep cattle, sheep, goats and poultry for food production and donkeys, mules and horses for transportation.[8]

The agricultural production of the peasants is used for the following main purposes: own consumption, paying tribute (taxes and rents), for sale at the market place (a network of daily and weekly markets covers the countryside), seeds for the coming agricultural season and participation in social life (ceremonial celebrations, membership in organizations like *mahaber, eder, dabo* etc.).[9]

It is beyond the scope of this study to further penetrate these aspects of peasant life. Instead, the analysis will turn to the pattern of land ownership and tenure.

6.2. Land Tenure

The Ethiopian land tenure system is extremely complex, different patterns being found between and within the provinces and even, sometimes, within *awradjas*. No detailed classification will be attempted here. For the following analysis, however, a rough classification of the main categories of landowners must be made. In Ethiopia, land can be owned

i) communally by peasant and pastoralist communities
ii) by official institutions (the Imperial Government, the Orthodox Church)
iii) by private persons
iv) by business companies

This classification does not claim to be exhaustive. It is though relevant for the situation in the 1960's and the early 1970's.[10]

Communal Ownership

The historical origin of this form of ownership has been discussed in Chapter 2. Here, some notes on the present situation will be added.

In the highland parts of Eritrea, Tigre, Beghemder and Godjam provinces, in Wag, Wadla Delanta and Lasta *awradjas* in Wollo province and in some *awradjas* in the northern part of Shoa province agricultural land is in the possession of peasant communities.[11]

Two main varieties of the communal form of land ownership are to be found.[12] In Amharic areas land is considered to be the property of all members of a kinship group which can claim ancestry to the same founding father (*akni abat*) of the area. The members of the living generation have use-rights to those parcels of land which they have inherited from their parents. But they cannot sell the plots which they cultivate since the actual ownership is vested in the founding father. A transfer of property rights outside the kinship group would violate the rights of coming

Map 7. COMMUNAL LANDOWNERSHIP

Source: Mesfin Wolde Mariam, *An Introductory Geography of Ethiopia* and Karl Johan Lundström, *Communal-Individual Land Ownership System in Northern Ethiopia.*

generations. The basic production unit in this system is the family, which has use rights to specified plots of land. These are inherited by the children. Pasture land is held in common.

In Eritrea and Tigre a form of land ownership called *diessa* is prevalent. Here ownership is vested in the village community. The land belonging to the village is distributed among its residents. At certain intervals (5 to 8 years) redistribution takes place in order to meet the needs of its new members. Redistribution is conducted by the village elders.

Earlier, there was a hierarchy of authorities, ranging from the local *gult* holder to the Abyssinian king, who claimed tribute from the peasants in the communally owned areas. The amount of tribute, which the peasants were obliged to pay, appears to have varied from place to place. Sometimes these payments approached those levied on the peasants in the conquered territories in the south. It was reported from Tschabi in Godjam that the peasants there in addition to tithe gave 1/5 of their harvest as well as occasional gifts of sheep to their *gult* holder. Furthermore they had to work for him.[13] This example shows that control over production is not necessarily dependent on juridical ownership of the means of production.

In 1966 *gult* privileges were officially abolished by law.[14] Those claiming tribute were thereby reduced to the Imperial Government and the

Orthodox Church. But this did not signify immediate changes in the social structure since the local *gult* holders were compensated by the Imperial Government.

> "In Gojjam the gultenyas [*gult* holders] have reportedly been made Government officials in every case, and the Ministry of Finance acknowledges that Gojjami gultenyas are rewarded with a proportion of the tax collected ... The local, social, political, and economic position of the former gultenya seems to be little affected. Gultenyas from other provinces have been compensated with grants of land in the southern provinces of Bale and Sidamo ..."[15]

Communal ownership in the village of Yebrage Hawariat in Godjam is described by a student:

> "The land holding system of Yebrage is traditional; characterized as belonging to all the descendants of the ancestors known as "Akni Abats". A few hundred years ago the first leading settler called "Konez" established a church, today known as Hawariat. He occupied the area in the name of the church. In the course of time other settlers, today referred to as Akni Abats, settled in Yebrage and established their right of landownership. As a result the land [was] divided into eight Akni Abats. This was the first stage of fragmentation. Later, the recognized children of the eight Akni Abats, known as "Menzer Abats", subdivided the land among themselves. The principle of father to children inheritance resulted in uncontrolable measure of parcellation and fragmentation. Though these Akni Abats may have died long ago, ownership is still under their names ... Apparently the land of Yebrage are registered as Church land and as Government land. But they have only Gult rights. Both of them don't have the right to confiscate or to use the land. So the absolute ownership of land rests in the hands of the settlers. They use the land communally and because of this there is a strong attachment to the land."[16]

The number of peasant family holdings in the communally owned areas was estimated to approximate 1 million in the mid 1960's.[17] Of these holdings some 95 % comprised 3 hectares or less and some 65 % comprised 1 hectare or less.[18] The inheritance pattern, the limited possibilities to increase the agricultural land and the population growth in recent times have all contributed to creating a complex pattern of fragmentation of the holdings. Holdings which were composed of 3 separate parcels or more (the maximum fragmentation was reported to be 9 parcels) constituted 47 % of all holdings in Godjam, 56 % in Beghemder and 59 % in Tigre.[19] Due to this fragmentation and the inheritance pattern many peasants must rent land from relatives or neighbours as a supplement to their own land. The following table shows the tenure pattern in the communnally owned areas.

TABLE 1. Percentage distribution of peasant holdings by tenure in the communally owned areas*

Province	Owned	Partly owned/ Partly rented	Rented	Total No. of Holdings
Tigre	75 %	18 %	7 %	253,000
Beghemder	85 %	6 %	9 %	190,000
Godjam	80 %	7 %	13 %	245,000
Wag *awradja*	76 %	23 %	1 %	30,000
Lasta *awradja*	56 %	38 %	6 %	54,000
Wadla Delanta *awradja*	76 %	23 %	1 %	30,000

Source: CSO, *Survey of Tigre;* CSO, *Survey of Beghemder;* CSO, *Survey of Godjam;* CSO, *Survey of Wollo.*

*Eritrea has not been included in CSO's sample surveys.

The category 'owned' in the table refers to ownership within the communal system. The proportion of wholly rented holdings indicates that tenancy exists in the communally owned areas, although only to a limited degree. The completely landless tenants are to a great extent composed of politically oppressed, economically exploited and socially despised minority groups. They are not allowed to own land. These groups make their living from handicrafts, supplemented by agriculture on a tenancy basis.[20]

The communal form of landownership has not been unchallenged. During the *mesafint* period, regional warlords attempted to appropriate land from the peasant communities. In the 1960's FAO experts renewed the attack and suggested that communal land rights should be transferred to individual land rights under the presumption that private inidividual land ownership was a prerequisite for agricultural development.[21] Recent attempts by the Imperial Government to measure the land in the communal areas were interpreted by the peasantry as a governmental plot to do away with their traditional system of tenure. The Government officials, who were to perform the land measurement, were met with hostility. Violent clashes occurred between armed peasants and Government troops, most recently in Godjam in 1968–69.[22]

The total area under communal ownership is decreasing. A transition to private ownership is taking place on the fringes of the region. Wealthy Addis Ababans are now buying up land and reducing the peasants to their tenants in those *awradjas* in northern Shoa which were in the 1960's still classified as having communally owned areas.[23]

An assessment of the communal landownership system must recognize that the prohibition to buy and sell land has prevented a sharp polarization of the peasantry into large scale owners and landless tenants. This does not, however, mean that all members of the peasant communities are equals. According to Hoben, political influence and power are crucial factors when it comes to a man's ability to obtain shares of fields in communities where he has *rist* claims.[24]

Another form of communal landholding is exercised by pastoralist communities. They include the different clans within the Beja, Saho, Afar, Somali and Oromo tribes who occupy lowland parts of Eritrea, Tigre, Wollo, Shoa, Hararge, Arussi, Bale and Sidamo provinces. These pastoralists depend almost exclusively on their livestock (camels, cattle and goats) and use large areas as grazing land. The Imperial Government does not, however, recognize them as legal land owners. Officially their land is classified as 'Government land'.[25] Estimates of the number of pastoralists in Ethiopia in the 1960's approximate 2 million[26]. The establishment of commercial plantations along the Awash river has seriously threatened the living conditions of the pastoralists there.[27]

The river valleys in western Ethiopia are sparsely populated by 'Prenilotic' groups, who also hold land communally for hunting and agriculture. Their land is also classified as 'Government land'.

The Imperial Government

The Imperial Government has developed from the Imperial Crown of the Manze dynasty. The relations between the Crown and the Government are intimate, and in many spheres there is no clear distinction between their roles and prerogatives. Such is the case in landmatters. A precise distinction between Imperial land and Government land has not been made.[28] In the following, this difficulty will be overlooked and the term Government land will be used except when it comes to pieces of land which are known to be in the possession of individual members of the Imperial family. The term Imperial land will then be used.

In 1969 a survey of Government land in units of 800 hectares or more was undertaken and presented in a report. According to this report, 42 % of the total area of Ethiopia was classified as Government land.[29] Since it is known that there are numerous pieces of Government land in units less than 800 hectares, it can be concluded that the Imperial Government was the owner of close to one half of the area of Ethiopia at the turn of the decade.

According to the report, 78 % of Government land had a "grazing only" potential and could not be used for agriculture without heavy investments in irrigation.[30] These areas were situated in the lowland parts of Hararge, Bale and Sidamo provinces, where 79 %, 81 % and 75 % respectively of these three provinces was classified as Government land.[31] It is these very regions that are the homes of the pastoralists mentioned above.

The bulk of the remaining Government land was classified as land with agricultural potential. It comprised 9.8 million hectares (19 % of the Government land) and was situated in the rainfed areas of Bale, Gemu Gofa, Illubabor, Kafa, Wollega, Shoa, Arussi, Wollo and Godjam provinces.[32] The report maintains that these lands are sparsely populated. This must be questioned. Other reports suggest that 'illegal' cultivation is frequent on Government lands classified as 'unutilized' (cf. Chapter 5).

The picture which emerges from the report is that the Imperial Government owns enormous areas of dry and barren land but limited areas of fertile agricultural land.

The historical origin of Government owned land is to be looked for in Menilek's conquests. Almost all Government land in units of more than 800 hectares is located in the southern provinces of Ethiopia.[33]

The total amount of Government land has been greatly reduced through Haile Selassie's land grant policies (cf. Chapter 5). The process of granting Government land to private persons has continued during the early 1970's. I have no figure for the total amount of land granted since the 1969 report was published. One example must suffice here. In Chilalo *awradja* 23,500 hectares of Government land was granted to private persons during the period 1962–69. Of this amount 1/3 was granted to landless people and 2/3 to already wealthy landed people. In 1971 alone 2,400 hectares were granted to persons who were already large landlords.[34]

The allocation of Government land to private persons is, to a limited degree, compensated for by the accumulation of land from individuals by the Imperial Government. This process goes under the name *gebretel*. It signifies that the Imperial Government confiscates the land of people who have failed to pay taxes for three consecutive years. *Gebretel* strikes in the first instance at poor owner-cultivators. In the 1950's and 1960's *gebretel* dispossessed several hundreds of owner-cultivators.[35]

The only way for a peasant to escape *gebretel* is to possess an official, written receipt showing that he has paid land tax (since 1968 agricultural income tax). The more remote areas in southern Ethiopia have been brought closer to Addis Ababa and Government machinery through highway construction programmes undertaken after world war II. The peasants and pastoralists which have inhabited these areas for generations have of course no receipts to show for the taxes paid. Much of their land has therefore been officially classified as 'Government land' and the inhabitants reduced to tenants or driven away.[36]

The Imperial Government has involved itself in commercial agricultural production. The two state farms in Awassa (Sidamo province) and Arba Minch (Gemu Gofa province) demonstrate the brutal effects that the official Ethiopian development policy inflicts on poor peasants and pastoralists. In Awassa, the Ministry of Community Development selected a large area for a 'development' project. A group of 400 pensioned soldiers were brought to the area. They chased away the Sidamo pastoralists from the land and set fire to Sidamo huts, in the belief that they would become the new proprietors. The area was thus cleared for cultivation. The Ministry of Community Development brought mechanized agricultural machinery to Awassa and established a state farm in the area. The pensioned soldiers, who had expected to be granted land ownership rights, were instead given employment as guards, servants and agricultural labourers on the state farm. Almost identical events preceded the establishment of the state 'development' farm in Arba Minch.[37]

The Imperial Government is also a shareholder in the large sugar and cotton plantations in the Awash valley which are controlled by multinational corporations.[38]

The conclusion which can be drawn from the land policies of the Imperial Government, as they have been pictured here and in Chapter 5, is that they benefit important private persons (who constitute a class of landlords), multinational corporations and the Government itself. The loosers are the poor peasants and pastoralists.

The Ethiopian Orthodox Church

The Orthodox Church traces its history to the fourth century, when the ruling dynasty in the Aksumite kingdom embraced Christianity. Since then the Church has been rewarded with land grants from the Crown in exchange for ideological support.

The Church presently exercises two principal forms of land control. The first form is direct ownership (*church gult*). Land under the direct administration of a church or a monastery is usually divided into plots and cultivated by the lower clergy. When a clergyman dies or otherwise terminates his office, the land he has held is given to his successor. The church-owned land functions as a salary for the lower clergy.[39]

The other form of Church land control is related to the traditional tribute rights held by Abyssinian authorities over peasant production. This form is called *samon* and it signifies that people whose land is registered as *samon* do not pay taxes to the Imperial Government but to the Orthodox Church. Among the holders of *samon* land are poor owner-cultivators as well as large landlords. *Samon* has been a popular form of tenure among the landlords since the tax rates of the Church are lower than those of the Imperial Government. In order to decrease their already insignificant tax burden, large landlords have been eager to register their land as *samon* land.[40]

An attempt to estimate the total area of Church land in Ethiopia is a difficult venture. The Church does not own "one half of the Empire of Ethiopia", which is sometimes suggested in newspaper reports. In the southern provinces Church land (*samon* + *church gult*) is of a moderate size. According to estimates made by the Ministry of Land Reform and Administration, *samon* + *church gult* amounted to less than 10 % of the measured area in all provinces in the southern half of Ethiopia, except in Shoa (14 %), Hararge (17 %) and Arussi (24 %).[41]

The bulk of Church land is found in the communally owned areas, i.e. in the regions which formerly constituted the Abyssinian kingdom. The numerous monasteries and churches here control vast areas of land, directly through *church gult* or indirectly through *samon*. No complete inventory of Church land, though, has come to my attention. Thus I refrain from speculating on how large the proportion of Church land is.

Private Persons

Private ownership of land prevails in the southern half of Ethiopia. It is estimated that approximately one half of the agricultural land in Shoa, Wollo, Wollega, Arussi, Hararge, Illubabor, Kafa, Sidamo, Gemu-Gofa and Bale provinces is under private ownership, the other half being Government land, Church land and land under tenures such as *maderia, siso* and *rist-gult**.[42]

The historical origin of private ownership must be viewed against Menilek's conquests and Haile Selassie's land grant Orders. Accordingly, there are great variations in the size of privately owned holdings. The freehold grants of one or one half *gasha* (40 resp. 20 hectares) given to landless persons, patriots and others have, through sales and inheritance, developed into smallholdings while the larger grants, especially those given by the Emperor personally, have remained large.

Private ownership of land in the southern provinces is connected with tenancy. Traditional farming techniques allow one family to cultivate a maximum of 10 hectares without using labour outside the family. Consequently, people who own larger areas than 10 hectares must lease their

* *Maderia, siso* and *rist-gult* are now being converted to privately owned land.

remaining land on tenancy basis if they want to keep it in production*. A preliminary classification of different categories of people involved in peasant agriculture can be set up accordingly:

Poor peasants
those operating the smallest holdings, i.e. up to 3 hectares.

Middle peasants
those operating from 3 hectares up to the maximum area which can be cultivated without using hired labour (10 hectares).

Rich peasants
those who themselves are cultivators but own additional land which they lease to tenants (10–20 hectares).

Landlords
people who own so much land that they do not have to practice agriculture themselves but can live on the rent from leased land (above 20 hectares).

It should be mentioned that this classification contains elements of arbitrariness and needs to be tested against the actual functioning of peasant society in southern Ethiopia.[43]

Tables 2 and 3 illustrate the tenure situation in the mid 1960's.

The tables show that an overwhelming majority of the peasants belongs to the poor peasantry. Only in Arussi and Shoa do the middle peasants account for somewhat larger proportions (20 % and 15 % respectively of the total). Since the tables bundle together all holdings comprising more than 5 hectares, it is impossible to distinguish between middle peasants and rich peasants.

The statistical surveys are based on peasant holdings. They do not supply information on the frequency of large landlords. The frequency of tenancy, however, gives an indirect clue to the existence of large landlords. A clear majority of the holdings in the provinces covered by the table is tenant holdings. Tenancy goes hand in hand with absentee ownership. Estimates from five provinces in the south (Shoa, Arussi, Wollega, Gemu Gofa and Sidamo) indicate that from 28 % to 45 % of the measured area in these provinces was held by absentee owners.[44]

All that can be concluded from this scanty data is that the bulk of the

* This holds only for the period prior to the introduction of tractors and agricultural machinery.

TABLE 2. Percentage Distribution of Peasant Holdings by Size.

Province	–0,5 ha	0,5-1 ha	1-2 ha	2-3 ha	3-4 ha	4-5 ha	5 ha–
Shoa	23 %	22 %	27 %	13 %	6 %	5 %	4 %
Wollega	29 %	36 %	24 %	7 %	2 %	1 %	1 %
Arussi	8 %	23 %	29 %	20 %	10 %	6 %	4 %
Hararge	45 %	31 %	16 %	5 %	1 %	1 %	1 %
Illubabor	32 %	37 %	21 %	3 %	2 %	–	5 %
Kafa	41 %	35 %	19 %	4 %	1 %	–	–
Gemu Gofa	73 %	19 %	6 %	2 %	–	–	–
Sidamo	73 %	18 %	7 %	1 %	–	–	1 %
Wollo*	55 %	25 %	14 %	4 %	2 %	–	–

Source: CSO *Survey of Shoa, Survey of Wollega, Survey of Arussi, Survey of Harage, Survey of Illubabor, Survey of Kafa, Survey of Gemu Gofa, Survey of Sidamo, Suryey of Wollo.*

TABLE 3. Percentage Distribution of Peasant Holdings by Tenure

Province	Owned	Rented	Partly owned/ partly rented	Total No. of holdings
Shoa	33 %	51 %	16 %	671,000
Wollega	41 %	54 %	5 %	191,000
Arussi	48 %	45 %	7 %	116,000
Hararge	46 %	49 %	5 %	263,000
Illubabor	25 %	73 %	2 %	112,000
Kafa	38 %	59 %	3 %	224,000
Gemu Gofa	53 %	43 %	4 %	114,000
Sidamo	61 %	37 %	2 %	376,000
Wollo*	55 %	24 %	21 %	290,000

*Wag, Lasta and Wadla Delanta *awradjas* are excluded.
Source: Same as for table 2.

cultivators are poor peasants, the middle and rich peasants forming a minority. The distinction between rich peasants and the lower stratum of the landlords is, or course, diffuse. The landlord category forms a continuum, the lower strata owning between 20 and 40 hectares and the upper strata owning thousands of hectares and forming the feudal aristocracy around the Imperial thorne. The apex of the landlord class is the members of the Imperial family who are the largest single owners of land in the empire.

Tenancy

Contemporary landlord-tenant relationships in the southern provinces have evolved out of the *gabar* system and reflect the subordinate position of the tenants. What follows is some general remarks on tenant obligations.

In general lease agreements between landowners and tenants are oral.

The agreements are not for specified periods of time but are valid until further notice. The lease can be terminated by either of the parties with immediate effect, rent being paid in kind or in cash. The most common form of paying rent is share-cropping, the share varying between 1/4 and 3/4 of the harvest. The proportions of the crop paid as rent is determined by the contributions of the landowner. If he contributes land only, the rent is generally 1/3 of the harvest, but if he also contributes farming implements (seeds, oxen, plough) the rent increases. Most of the tenants supply their own oxen, ploughs and seeds and, consequently, the most common rent is 1/3 of the crop. This goes under the name *siso arash*. Payment of a fixed amount of cash or agricultural products is widespread in areas where cash crops are grown.[45] There is also a form of tenancy which is like a master-servant relationship. Here, the landowner allots a piece of land to a landless person, who in turn cultivates it and keeps the whole yield. In return he works the fields of his landlord (together with others), accompanies him to the market and performs other servant functions.

Rent is paid at harvest time. Rich peasants and small landlords collect the rent themselves while large landlords, who do not reside in the area, collect rent through local representatives.

In addition to rent, tenants in many places pay tithe (*asrat*) to the landowner. Until 1967 tithe was a tax payable to the Government. The praxis was that landowners paid the tax to the Government and as compensation collected 1/10 of the total harvest from their tenants before other tributes were collected. In 1967 Parliament abolished the tithe but information about this legislation failed to reach the tenants in many areas. They thus continue to pay tithe, which since that year is retained by the landowners.[46]

Furthermore tenants traditionally provide personal labour for their landowner. They cultivate the fields which are used for the consumption of the landowner's household. On holidays tenants show their respect by presenting gifts (chickens, eggs, etc.) to the landowner.

Tenants are not, as a rule, compensated for the permanent improvements they have made on their holdings. On the contrary, if a tenant increases the value of the holding by planting trees, digging ditches etc., the landowner often responds by increasing the rent.

The obligations described above are of course not valid for all tenants. There are considerable differences between provinces and even between *awradjas*. In specific cases the relations between landowners and tenants vary from close co-operation and readiness to help each other to extreme exploitation.

The following hypothesis can be made: Monopoly or oligopoly ownership over an agricultural area goes together with an unstable situation for the tenants in the area, while in an area where the ownership of land is spread among many small owners the situation for the tenants in the area is more stable. Even if the assertion is not scientifically indisputable, it is worth a discussion.

Persons who own smaller areas, say 10 to 20 hectares, and who do not hold positions within the state administrative apparatus have to practice agriculture for their own subsistence. They live close to their holdings and co-operate with the tenants who cultivate their additional land. The relationship between owners and tenants is direct and personal. In many areas they share a common language and cultural tradition and there may also be blood relationships between them. It is obvious that tenants on the land of such owners are in secure possession of their holdings.

If a person owns several *gashas* of land and his land is divided into several small tenant holdings, it is impossible for him personally to supervise the agricultural operations, and above all, the collection of rent. Furthermore, such a large landlord is usually an urban resident, holding a governmental office. So, in each *woreda* where he owns land, he must employ a local representative to look after his economic interests. A landowner of this magnitude has probably received the land as an Imperial grant (or his parents have received the grant) so he is personally a stranger in the area where he owns land. His relations to the tenants are indirect and impersonal. The local representative of the owner is for his own living dependent on the tenants paying their rent punctually. The representative cannot be too flexible since his fitness for the job is measured in his capability to look after the interests of his employer. The representative may himself be a stranger in the village he supervises. In such a situation, it is not difficult to imagine that the landlord's superiority, being built into the tenancy relationships, is utilized to one-sidedly satisfy the owner's economic interest without being responsive to the tenants' needs. The owners are only interested in the tribute they can collect from the farmers and their local representatives is their instrument.

In conclusion, the tenancy relationships imply surplus appropriation, the consequence of which is often to make the tenants apathetic to innovation.

Viewed from another angle, the conditions of tenancy are a prerequisite for the power and wealth of the landlords. The structure has built-in contradictions.

Notes to Chapter 6

[1] CSO, *Statistical Abstract 1971,* p. 11.

[2] *Ibid.,* p. 41.

[3] *Ibid.,* pp. 41 ff.

[4] *Ibid.,* p. 11; *TFYDP,* p. 183.

[5] Mesfin Wolde Mariam, *An Introductory Geography of Ethiopia* (1972), pp. 90–96; H.P. Huffnagel, *Agriculture in Ethiopia* (1961), pp. 140–165.

[6] Information obtained from EPID's headquarters.

[7] *Statistical Abstract 1971,* pp. 40 f.; Mesfin Wolde Mariam (1972), *op. cit.,* p. 100. The figures given in the text must be considered to be very rough indicators. Actual yields vary considerably from year to year and from province to province due to different ecological conditions and a multitude of local crop varieties.

[8] *Statistical Abstract 1971* gives the following estimates of livestock in Ethiopia: cattle, 26 million; sheep, 13 million; goats, 11 million; donkeys, 4 million; mules, 1 million; poultry, 48 million. A considerable proportion of these animals were not held by peasants but by pastoralists.

[9] For a detailed account of peasant organizations and their functions see Arne Lexander, *The Changing Rural Society in Arussi Land* (1968).

[10] For detailed classifications of the forms of tenure see for example: Tamrat Kebede, Land Tenure and Agricultural Development in Ethiopia, in *Peace Corps Ethiopia Training Programme* (1970); Worku Moges, *Land Reform in Relation to Agricultural Development in Ethiopia* (1968); Pankhurst (1966), *op. cit.;* Mesfin Wolde Mariam (1972), *op. cit.*

[11] Mesfin Wolde Mariam (1972), *op. cit.,* p. 85.

[12] For detailed accounts of the different types of communal landholding in Eritrea, Tigre Beghemder and Godjam, see for example Ambaye Zekarias, *Land Tenure in Eritrea (Ethiopia)* (1966), Dan Franz Bauer, *Land, Leadership and Legitimacy among the Inderta Tigray of Ethiopia* (1972), Hoben, *op. cit.,* Wolfgang Kuls, *Bevölkerung, Siedlung und Landwirtschaft im Hochland von Godjam* (1963).

[13] Kuls, *op. cit.,* pp. 22 f.

[14] Negarit Gazeta, *Proclamation No. 230 of 1966. A Proclamation to Amend the Land Tax Proclamation of 1944.*

[15] Dunning, *op. cit.,* p. 283.

[16] Dillnessa Zewdie, *Land Use Study, Ten Selected Farmers in Yebrage Hawariat* (1971), pp. 19 f.

[17] Information obtained from CSO.

[18] CSO, *Report on a Survey of Tigre Province* (1967), p. 25 (hereafter: *Survey of Tigre);* CSO, *Report on a Survey of Beghemder Province* (1968), p. 33 (hereafter: *Survey of Beghemder)*; CSO, *Report on a Survey of Godjam Province* (1966), p. 25 (hereafter: *Survey of Godjam); CSO, Report on a Survey of Wollo Province* (1967), p. 32 (hereafter: *Survey of Wollo*).

[19] *Survey of Godjam,* p. 25; *Survey of Beghemder,* p. 35; *Survey of Tigre,* p. 26.

[20] Lawrence–Mann, *op. cit.,* p. 332.

[21] *Ibid.,* pp. 286, 316.

[22] For accounts of Godjami resistance to Government interference in land taxation and land tenure procedures see Gizachew Adamu, *A Historical Survey of Taxation in Godjam, 1901–1969* (1971) and Schwab, *op. cit.*

[23] Information obtained orally.

[24] Hoben, *op. cit.,* pp. 85 f.

[25] The pastoralists' legal rights to their pastures have been under debate in the Ministry of Land Reform and Administration. See Desta Asfaw, Tesfaye Wolde Mariam, Wondwossen Hailu and Zemedhun Wolde Gabriel, *Nomadic Areas* (1973).

[26] Information obtained from CSO.

[27] For an account of the destructive effects of commercial plantations on the living conditions of the pastoralists in the Awash valley, see Bondestam (1973), *op. cit.*

[28] Information obtained from the Ministry of Land Reform and Administration.

[29] *A Policy Oriented Study*... Vol. II, pp. 11 ff., 18.

[30] *Ibid.*

[31] *Ibid.*

[32] *Ibid.*

[33] *Ibid.* In Godjam Government land amounts to as much as 21 % of the province. But Government land here is concentrated to the Blue Nile gorge and the lowlands bordering Sudan.

[34] Haile Wolde, *Problems of Land Tenure in the Implementation of CADU Project* (1973), p. 24.

[35] Haile Menkerios, The Present System of Land Tenure in Ethiopia, in *Challenge,* Vol. X, No. 2, 1970, p. 20.

[36] Information obtained orally.

[37] Siegfried Pausewang, Peasant Society and Development in Ethiopia, in *Sociologia Ruralis,* Vol. XIII, No. 2, 1973, pp. 180 f.

[38] For details see Bondestam (1974), *op. cit.*

[39] *Report on Land Tenure Survey of Welega Province,* pp. 1 f.

[40] *Ibid.*

[41] Yoseph Haile, *op. cit.,* Appendix I–IV.

[42] *Ibid.*

[43] For a classification of different categories of peasants and landlords adapted to the conditions in Arussi province see Henock Kifle, *An Analysis of the CADU Credit Programme 1968/69–1970/71 and its Impact on Income Distribution* (1971).

[44] T.C. Varghese, Land Tenure Analysis: Research Achievements and Further Areas to be Studied, in *Seminar Proceedings on Agrarian Reform,* p. 74.

[45] Information compiled from CSO, *Report on a Survey of Shoa Province* (1966) (hereafter: *Survey of Shoa*); CSO, *Report on a Survey of Wollega Province* (1967) (hereafter: *Survey of Wollega*); CSO, *Report on a Survey of Arussi Province* (1966) (hereafter: *Survey of Arussi*); CSO, *Report on a Survey of Hararge Province* (1968) (hereafter: *Survey of Hararge*); CSO, *Report on a Survey of Illubabor Province* (1968) (hereafter: *Survey of Illubabor*); CSO, *Report on a Survey of Kafa Province* (1968) (hereafter: *Survey of Kafa*); CSO, *Report on a Survey of Gemu Gofa Province* (1967) (hereafter: *Survey of Gemu Gofa*); CSO, *Report on a Survey of Sidamo Province* (1968) (hereafter: *Survey of Sidamo*).

[46] Schwab, *op. cit.,* pp. 120 ff.

Chapter 7 Attempts at Promoting Peasant Agriculture

7.1. The Package Approach and the Package Projects

In Ethiopia the attempts at transforming peasant agriculture have utilized the package approach (cf. chapter 5). According to TFYDP, "package programmes must be seen as part of a long, slow-experimental and gradual diffusion process of the transformation of traditional peasant agriculture.[1] The idea of the package approach has, as its point of departure, a situation characterized by scarcity of resources for development purposes. Therefore, agricultural development can be accelerated through a concentration of inputs and activities to geographically delimited regions. Here, a locally well-coordinated attack, consisting of a number of measures forming a 'package', is launched upon "the most important factors which are preventing development".[2] Initially, this will mean an increased inequality between those regions where package projects are launched and those which are left on their own. In the long run, however, it is supposed that the concentration of inputs will pay off in the form of experience, innovations, production growth and increased tax revenues. These outputs can later be used for investments in new regions.[3]

Up to 1973 the main manifestations of the package approach to developing peasant agriculture were the WADU project[4], the CADU project and the minimum package programme. CADU and the minimum package programme affect peasants throughout Ethiopia. Their organization, activities and goals will be outlined below.

The first comprehensive package project to be established in Ethiopia was the Chilalo Agricultural Development Unit – CADU. The project, jointly financed by the Imperial Government and the Swedish Government, started its activities in 1967. The first project period covered the years 1967–70 and the second period the years 1971–75. In 1975 the Swedish Government will decide whether to continue its support of CADU for the period 1975–80. For the years between 1967 and 1975 the total expenses for CADU will amount to some Eth. $ 43 million, of which the Swedish Government pays approximately two-thirds, this sum being part of the official Swedish development assistance to Ethiopia.[5]

CADU operates in Chilalo *awradja* in Arussi province, but experiences gained from CADU are diffused throughout Ethiopia's agricultural regions through the minimum package programme. The first nine minimum package projects were established in 1971, and since then nine to ten new projects have been launched annually. The minimum package programme is jointly financed by the Imperial Government, the Swedish Government (SIDA) and the World Bank Group.* Total costs for the period 1971–75 are estimated to amount to approximately Eth. $ 28 million. These three contributors will furnish roughly one-third each.[6]

CADU and the minimum package projects are administered through a special department within the Ethiopian Ministry of Agriculture. The department is called the Extension and Project Implementation Department–EPID. Many of EPID's top executives are expatriate experts, notably Swedes.

The activities which form the CADU package include agronomic research, dissemination of research results, provision of agricultural inputs, a credit programme, a marketing programme, improvement of infrastructure, vocational education and a co-operative promotion programme.

The components of the package are linked to each other in the following manner.[7] *Research*, which mainly takes place on the project's large mechanised farm in Kulumsa, is necessary for the identification of major agronomic problems. Research is being done of fertilizer trials, variety trials, seed tests, crop protection trials, soil tests, livestock improvement, forestry and agricultural implements. The results of this research are then put into practical use to improve agricultural production (including livestock production) quantitatively and qualitatively. An *extension service* is established in order to communicate the information from the project to the peasants. The project region is subdivided into extension areas where agricultural extension agents and model farmers demonstrate the effects of new agricultural techniques (in particular fertilizing). The extension agents cultivate demonstration plots and field days are held. The model farmers, selected by the peasants in the neighbourhood, are provided with inputs by the project and are instructed by the agents. The results of these experiments are supposed to give the peasants an opportunity to compare the difference in yield between their own traditional methods and the technique which the project recommends. The new *inputs* provided by the project include fertilizer, improved seeds and improved farm implements adapted tc the man and animal based technology. The inputs do not include tractor

* In addition to these three main contributors minor contributors include USAID (United States Agency for International Development), FAO (Food & Agricultural Organization of the United Nations), DANIDA (Danish International Development Agency) and ODA (Overseas Development Authority of the United Kingdom).

operated machinery. In order to make it possible for the peasants to utilize the recommended inputs, the project provides a *credit programme* to finance the peasants' acquisition of fertilizer etc. The peasants must be interested in increasing their yields. The project provides them with an economic incitament: the opportunity to dispose of their surplus production on a *cash market*. The project's marketing division buys the surplus brought about by the new agricultural inputs and sells it on the national market. Local *roads* are built to enable the transportation of crops from the fields to project trade centres.

The credit programme and marketing activities play a key role in project philosophy. They open the door for peasant participation in project activities while binding them to a money economy (credits must be repaid in cash). The peasants' connection to the project is institutionalised through *co-operative societies*. The peasants participating in the credit programme are encouraged to form co-operatives which are to perform the credit and marketing operations. Gradually the co-operatives merge into secondary co-operative unions which are supposed to take over all project activities and the whole development effort in the region.

These activities are supposed to generate 'agricultural development'. This term is, however, vague and can be interpreted in many ways. One interpretation is provided by the major financing body of the package projects, the Swedish Government.

As part of the official Swedish development assistance, the goals of the package projects have one of their points of departure in the officially stated principles for the Swedish development assistance. These principles were laid down by the Swedish parliament in 1962 and 1968. According to the Government bills, the goal of Swedish development assistance is to improve the standard of living for poor peoples; the motive for development assistance is feelings of international solidarity.[8] The social and political aspects of the assistance are pointed out explicitly:

> "It is difficult to set out social and political goals for development assistance. It cannot be taken for granted that those social and political systems and principles to which we adhere are relevant or attainable for all developing countries. Nevertheless it may be reasonable to try to direct the development assistance so that it, as far as it is possible to judge, contributes to a development of society in the direction towards political democracy and social equality. It is not in accordance with either the goals or the motives of the Swedish development assistance, that it should contribute to a conservation of an antiprogressive social structure."[9]

In the Swedish development assistance programmes agricultural development is a main theme, and CADU and the minimum package programme form the most comprehensive Swedish-sponsored undertaking in this field.

In the preparatory phase of CADU the following main objectives were suggested:

> "a) to improve the ability of the local population to participate in and be responsible for the local administration and development effort and thereby improve the economic and social conditions in the area;
>
> b) to create possibilities for an expansion of the programme."[10]

The importance of popular participation was enhanced:

> "With respect to the first target it must be underlined that the project not only aims at an increased production. This could probably be achieved most easily through large scale farming and big industrial ventures, often under foreign management. The more important aspect of this project is, however, to develop the ability of the local people to deal with their own problems and to competently lead the progress of their society. This may perhaps be a slower but more secure and long-lasting approach, which will encompass and benefit the whole population and not just a fraction of it."[11]

This is all in accordance with the Swedish overall goal for development assistance. It is the emphasis on the local people (= poor peasants) which officially justifies a Swedish involvement in Ethiopia.

The goal description for CADU during the first project period (1967–70) states:

> "The success of the three-year programme will mean the initiation of social and economic development in the Project area. This shall essentially be achieved by improving the ability of the local population to participate in and eventually be responsible for the development effort and the local administration. The active participation of the local population in the project is not only a means of development but is regarded as a goal in itself."[12]

The inevitable conclusion of this statement is that CADU aims at a virtual peasant mobilization, which has clearly political implications. It follows as a practical corollary that the aim is popular influence over and an eventual take over of the decision making process at the local level (*woreda* and *awradja*). But this aim is incompatible with the principles and practise of Imperial rule, which at the local level is embodied in the governor system.

It may seem paradoxical that the Imperial Government supports a project which nourishes ideas about political mobilization of the masses, ideas which in the long run must undermine Imperial authority. I suggest that the paradox is apparent, and that the Imperial Government never intended to let the project activities be dominated by political mobilization of the peasants. It was suggested in chapter 5 that one of the motives for the Imperial Government to invite foreign technical assistance was to speed up production in Ethiopia. CADU can be viewed in this way. The important thing for the Government was the agronomic research and better agricultural practices introduced by CADU which makes it possible to increase agricultural production greatly. The Government paid lip service to the goal of popular participation, since the foreign assistance agency insists on it.[13]

7.2. Experiences from Chilalo

The key elements in CADU's package which have direct relevance for the peasants are the credit programme and the co-operative promotion programme.

A Brief Review of CADU's Credit Programme

CADU gives credits to farmers for purchases of fertilizer and seeds. The credits are given in kind. The repayment period is 12 months, the interest rate is 12 %, and 25 % of the credit value must be paid in advance.[14] When the credit programme started in 1968 everybody involved in agriculture was welcome to participate. Tenants had, however, to present a written lease agreement on their holding signed by the landlord and drawn up in a form which was agreeable to CADU.[15] This regulation was based on security considerations, a tenant without a written lease might be evicted and hence, unable to repay the credit.

TABLE 4. CADU's credit programme 1968–1973.

Year	Number of credit-takers	Total value of credits	Average amount of credit per credit-taker
1968	189	15,700 Eth. $	85 Eth. $
1969	868	158,500 "	183 "
1970	4,769	502,900 "	105 "
1971	14,164	1,437,500 "	102 "
1972	12,624	1,124,000 "	89 "
1973	13,302	958,500 "	72 "

Source: *CADU Project Description 1971; CADU Annual Reports 1971/72 & 1972/73.*

TABLE 5. Categories of credit takers in CADU's credit programme and their shares in the credit volume (figures are rounded off).

	1968		1969		1970		1971		1972		1973	
	% of credit-takers	% of credit-volume	% of credit-takers	% of credit-volume	% of credit-takers	% of credit-volume	% of credit-takers	% of credit-volume	% of credit-takers	% of credit-volume	% of credit-takers	% of credit volume
Tenants	9	4	15	5	32	28	39	36	30	30	22	Information not available
Landowners 1–10 ha.	43	17	51	25	40	39	59	57	62	56	69	
Landowners 11–20 ha.	19	19	17	20	9	14	3	5	7	10	8	
Landowners 21–40 ha.	11	24	8	12	4	6	1	2	2	4	2	
Landowners over 40 ha.	10	33	6	35	1	2	–	–	–	–	–	
Unknown status	9	4	4	4	14	12	–	–	–	–	–	

Source: Henock Kifle, *An Analysis of the CADU Credit Programme 1968–69 and 1970–71 and its Impact on Income Distribution* (1971); Michael Beyene, *An Analysis of CADU Credit Programme 1971/72 – 1972/73* (1973); SIDA, Utredningsbyrån.

Before the advent of CADU local money lenders and relatives were the only credit institutions available for the peasants in Chilalo. The interest rate charged by the money lenders stayed between 50 and 100 percent.[16] By comparison, CADU was a much better source of credit for the peasants, and the project made sure that the credits were used for agricultural purposes by giving them in kind. The expansion of the credit programme is shown in table 4.

The total number of peasant holdings in Chilalo was estimated at some 60,000 in the mid 1960's; of these some 50 % were tenant holdings.[17] This means that CADU's credit programme reached close to 1/4 of Chilalo's peasants in 1971.

Fertilizer has mainly been used in wheat production and the surplus production has been sold. CADU's marketing division has bought wheat from the participants in the credit programme at a fixed price. According to CADU's figures the yield of wheat on peasant holdings has increased between 50 % and 100 %, due to fertilizing. The net return in cash for the participants has been Eth. $ 125 per cultivated hectare* (1969).[18]

The different categories of credit takers and their respective shares in the credit volume are shown in table 5.

In addition to the credits, CADU's marketing division sold fertilizer in cash. The cash buyers numbered close to 4,000 in 1973.[19] It goes without saying that the cash buyers were relatively wealthy individuals.

Concerning the credit programme, the picture was one of rapid expansion in participation during the first four years followed by stagnation since 1971. Both the expansion and the subsequent stagnation can be interpreted in terms of wheat prices. From 1968 to 1971 prices were relatively high and invited a commercialization of peasant agriculture where credits were available. Since then wheat (and other grains) prices have dropped (the 1972 prices were extremely low) and have made peasants more reluctant to take credit.[20]

The distribution of the credits show that owner-cultivators ('landowners 1—10 ha.' in the table) every year have constituted the largest credit taking group. Within this group, owners of more than 3 hectares have dominated.[21] But when it comes to the credit volume, one finds that during the first two credit seasons the landlords, although numerically few, dominated the credit programme, they appropriated close to one half of the credit volume, while the tenants' share of the credits was negligible. This

* Benefits are calculated to be the difference between income from sale of the surplus production at prevailing market prices and repayment of fertilizer credits. These kinds of calculations are discussed in chapter 10.

tendency was incompatible with the main goal of CADU, the credit programme reached its main target group only marginally. CADU responded by issuing a regulation according to which large landlords (defined as people who cultivate more than 20 hectares) were not allowed to take credits. But they were still allowed to buy fertilizer in cash.[22] From the table above it can be seen that the regulation had effect in that the proportion of participants holding more than 20 hectares has declined drastically since 1970 although they have increased in absolute numbers. Instead more tenants have participated. But still, due to the share-cropping system, the landlords can benefit indirectly from the credit programme:

> "Depending on the tenancy relationship from one third to a half of this increased income of the tenants has gone to the landlords. Thus, as long as share-cropping exists, it will not be possible to bring about a more even distribution of income between these two groups of farmers; for as the income of the tenants increase (due to the Project's activities) the landlords' income will also increase by a certain proportion".[23]

Furthermore, it has been reported that the introduction of fertilizer and the general agricultural boom in Chilalo have resulted in a gradual shift of the share-cropping arrangements from *siso arash* (1/3 of the harvest is paid as rent by the tenant) to *ekul arash* (1/2 of the harvest is paid as rent by the tenant).[24]

The conclusion is, then, that the credit programme cannot function as an instrument creating social and economic equality in Chilalo as long as land is unevenly distributed and the share-cropping system prevails.[25] These two factors are manifestations of the basic social structure in Chilalo (and in southern Ethiopia in general).

And to generate basic structural change is beyond the capacity of a project which is administered through a Government whose power is dependent on shoring up the prevailing social structure.

On the other hand, it should not be overlooked that a number of peasants, due to the credit programme, have got increased cash incomes. But it is difficult to evaluate this. Prices on manufactured consumer goods have also been on the increase for a number of years and no study has yet been made on how cash incomes, generated through the credit programme, have been used.

A Brief Review of CADU's Co-operative Promotion Programme

If one judges from the goal description, the promotion of peasant co-operatives, being the embodiment of the goal 'popular participation', should have top priority in the project activities. Reality shows, however, something else. During the project period 1967–70 little was reported from the co-operative front. Activities aimed at forming co-operative societies are not mentioned in CADU's annual reports until 1969/70. The project management played a waiting game and no co-operative societies were formed. Attempts at organizing another platform for popular participation – the Awradja Development Committee – was also a failure. The Development Committee was to consist of representatives of Chilalo's peasants, local Government officials and CADU staff. The committee was meant to function as a forum for discussion and coordination of the various development efforts in Chilalo.[26] The committee held only a few meetings which were failures from the point of view of popular participation, since the peasants did not dare to voice their problems in front of the Government officials and landlords in attendance. Therefore, the project management dropped the whole idea.[27] At the end of the 1967–70 period the project manager suggested that in the future "the achievement of economic development should be the ultimate goal with which the others should not be allowed to interfere."[28]

This passive, or even negative, attitude on the part of the project management in giving substance to popular participation was critisized in a report written on CADU's 1967–70 period by an independent project evaluation team. The report stated that "the project is not likely to contribute significantly to economic development unless the active participation of the local population is obtained".[29]

At the beginning of the 1971–75 CADU project period, efforts at co-operative promotion were made. The project personnel summoned meetings with the peasants within the project region and lectured on the advantages of co-operative societies. As a result a few co-operatives were formed. In 1973 there were four co-operative societies in various stages of promotion in Chilalo. Together their membership amounted to 2,107.[30] The societies operated along commercial lines, their main functions were to handle the credit applications of members, distribute fertilizer to members, construct storage facilities, collect members' surplus production of wheat and sell it to CADU's marketing division and to distribute profits to members.[31] A latent political potential in the co-operatives was indicated by the society in Bilalo, which used its profits to purchase land and distribute it to members who were tenants.[32]

It must still be concluded that CADU's co-operative promotion up to 1973 was of a limited nature and in no case did it reach the stage of mass mobilization. The reasons for this must be sought in three main directions, i) the negative attitude on the part of the Imperial Government and the local Government officials to the idea of encouraging genuine peasant organizations,[33] ii) the probability of a technocratic bias in the outlook of the project management where Swedish agronomists and economists held many influential positions[34] and iii) the deeply rooted and, indeed, justified suspicion among the peasants against Government projects.

The Growth of Mechanized Farming

When CADU started its activities in 1967 there were only a few commercial farms using a mechanized technology in Chilalo.[35] Since then the number of commercial mechanized farms has increased rapidly; Chilalo has become one of the centres for large scale commercial production of cereals in Ethiopia.

In 1972 there were altogether 126 commercial farmers in Chilalo. They cultivated about 23,000 hectares which is 10 % of the total cultivated area in Chilalo.[36] The farmers had the use of 184 tractors and 37 combines. The average cultivated area per farmer was 183 hectares (approximately 4.5 *gashas*). Actually, 85 % of the commercial farmers cultivated less than 200 hectares.[37] Local landlords held 56 % of the mechanized area while 44 % was held by contractors–financially strong persons from outside the region who had contracted land from landlords.[38] The main cash crop of the commercial farms has been wheat, which has been sold on the market in Addis Ababa.

The introduction of mechanized commercial farming has not left peasant agriculture untouched. In 1971 a detailed investigation on how the mechanized farms have affected the living conditions of the poor peasantry was made. The investigation concerned five extension areas within CADU's area of activity. The main findings of the investigation are summarized below.[39]

The investigation shows that mechanized farming in the area under study is of relatively recent origin. Before 1968 only a negligible part of the area was under mechanized cultivation, but since that year mechanization has been expanding at an increasing rate. In 1970 about 25 % of the tractor-cultivable area was under tractor cultivation. During the period 1968–70 mechanization prodeeded fastest on lands belonging to persons owning at least 20 hectares (holdings larger than 20 hectares comprised 36 % of the surveyed area). In all, 9,700 hectares were under tractor-cultivation in 1970.

In 1969 and 1970 a total of 538 tenants moved out of the area under study and among them 392 families were evicted as a direct consequence of mechanization. These figures were given by *golmasas*, a kind of official whose duty it is to keep records of people moving in and out of their area of responsibility. The actual number of tenants who had left the area seems to have been still greater. The *golmasas* in many cases mentioned that more tenants than the number indicated by them had left. The author of the investigation calculated that the true number of evicted tenants in 1969 and 1970, due to mechanization, may have been between 450 and 500. The total number of tenants in the surveyed area in 1968/69 was estimated at 3,420. Thus, the tenants evicted in 1969 and 1970 amounted to between 13.2 % and 14.6 % of the 1968 tenant population. If previous years are included, approximately 20 % of the tenant population may have been evicted. The tenants were usually told to leave after the harvest was brought in.

According to the investigation, some of the evicted tenants had moved to nearby areas but the majority had left Arussi province. It was mentioned that the tenants had left for places in southern Shoa, Bale and Hararge provinces, believing and hoping that in those areas there was land available for cultivation.

Eviction is not the only negative consequence of mechanization for small farmers. The price of land is increasing due to mechanization. This means that tenants who are not yet evicted face higher rent rates.

Mechanization does not only affect agricultural land. Grazing land is also being mechanized, which means a decrease in land available for cattle-breeding. According to the investigation, this phenomenon has negative consequences for poor peasants in general (both tenants and owner-cultivators). For the peasants, cattle-breeding is an important complement to crop-growing. In times of crop failure livestock functions as an economic reserve. The small owner-cultivators do not possess enough land for both agriculture and grazing. Therefore, the usual practice has been to rent pasture land from large landowners. The introduction of mechanized farming on former grazing land means, then, that the poor peasants either have to specialize in crop production alone, or have to move to the areas where they can continue their traditional practises. The poor peasants are reluctant to give up cattle-breeding because of economic insecurity, and some of them can be expected to leave the area. According to the investigation, some owner-cultivators had already left.

On the basis of these findings, the investigation predicted that if mechanization continues to expand, it would mean the breaking up of the traditional peasant society. An increasing part of the tenant population will be evicted. The investigation predicted that if the trend continues, 68 % of the 1968 tenant population in the surveyed area will have been evicted by 1975 and more owner-cultivators will voluntarily leave the area because of the limited possibilities for cattle-breeding. Thus, the progressive depopulation of the area may, at a later stage, in itself act as a catalyst for the remaining poor peasants to leave the region.

The prospects of the investigation referred to above seem to be on its way to fulfilment. Although it is impossible to arrive at an exact figure on the number of evicted tenants in Chilalo, some approximations can be made. An estimate of Ethaya in the north of Chilalo showed that the tenant population there decreased from 46 % of the peasant population in 1968 to 12 % in 1972.[40] Ethaya is one of the places where mechanization has proceeded at great length. Estimates for all Chilalo indicate that the number of peasant holdings in the *awradja* decreased by 2,300 during the period 1966–1972.[41] It is not far-fetched to suggest that most of these were tenant households which were victimized by the process of mechanization.

The general reason for the mechanization process must be looked for in the Government policy of encouraging mechanization by subsidies. In the particular case of Chilalo it is beyond doubt that CADU also has been involved in the mechanization/eviction process. Firstly, CADU's large mechanized research farm in Kulumsa must have been instrumental in transmitting the idea of the profitability of commercial agriculture to landlords in the area. Secondly, it is a fact that CADU in the beginning provided expert advice to the few commercial farms that were already in existence in Chilalo at the advent of the project.[42] Thirdly, it is also a fact that much of the fertilizer provided on credit by CADU in 1968 and 1969 reached large and medium landlords.

It can be argued that CADU's contribution to large scale mechanized farming was of an indirect nature and not consciously intended. The interesting thing to be noted for the analysis presented in the subsequent chapters is, however, that the policy of simultaneously encouraging both large scale commercial agriculture and peasant agriculture generated contradictions. Large scale commercial farming could advance only at the cost of peasant agriculture.

The experiences from Chilalo show that policies aimed at promoting peasant agriculture, however well-intended they might be, are bound to be deficient as long as they do not encompass the political system of which the peasantry is a subordinated part.

7.3. The Minimum Package Projects

Goals

The minimum package projects represent the realization of CADU's second main goal "to communicate the most important experiences gained by CADU to peasants throughout Ethiopia". The minimum package programme was launched in 1971 and coincided with CADU's second project period. The goal description for the minimum package projects and for CADU for the period 1971–75 are almost identical. The first main goal of EPID concerning the minimum package projects is:

> "The first main goal shall be the achievement of an improved standard of living for the farm population. The activities towards this end shall be so conducted as to ensure the participation of the population in and their assuming of increasing responsibility for those activities.
>
> EPID shall endeavour to avoid adverse employment effects and to observe opportunity to create additional farming possibilities. The activities shall be directed mainly towards farmers in the lower income brackets."[43]

The target group of the project activities is the same as for CADU—the poor peasants. Compared to the formulation of goals for CADU's first period, the present formulation is more vague, but it can still be interpreted as outspoken support for popular participation.

The third sentence in the goal description is conditioned by CADU's negative experiences of mechanization in the project area. It is the eviction of tenants that is referred to. The planners behind the minimum package projects realize that large scale commercialization/mechanization is a phenomenon that is not always consistent with the interests of smallholders. This is mentioned in the Master Plan for the projects' activities.

> "The availability of new agricultural techniques has also encouraged large land-owners to take up farming themselves and to evict some of their tenants".[44]

The discussion is elaborated like this:

> "Although tractorization still may prove attractive to the individual it will lead to eviction of tenants or labourers. It will presently be impossible to find alternative employment possibilities for this labour if such eviction should take place on a large scale. From the society's point of view the tractorization has in this case resulted in a substitution of scarce foreign exchange for plentiful labour."[45]

The collision course between large scale commercial agriculture and development efforts directed to subsistence farmers was known to the planners. In order to prevent the negative experiences of CADU from being repeated in the minimum package project areas a number of measures were recommended. Among them were:

i) The minimum package areas and CADU's project area will be given priority in the implementation of future legislation on tenancy relationships.[46] The reforms are contained in a draft Agricultural Tenancy Relationships Proclamation which, at the time of publishing this book, was still pending in the Ethiopian Parliament. The draft proclamation has, however, its limitations—a fact that the development planners are aware of. "The draft tenancy law when implemented may check this tendency [eviction], but in no way prevent it".[47]

ii) It was therefore suggested that the Imperial Government should make an effort to reserve cultivable Government land for evicted tenants.[48]

iii) Furthermore the Plan of Operation suggested that the minimum package areas should have priority in the implementation of future legislation on local self administration.[49]

The above mentioned measures are enumerated in the Plan of Operation under the heading "supporting measures". It is also stressed that they do *not* fall under the responsibility of EPID. The reforms can only be undertaken by the Imperial Government (some of the reforms in collaboration with the Parliament).[50]

This is a matter of course, but it is important to draw attention to it. The political pre-requisites for the successful performance of the minimum package projects are in the hands of a Government which is based upon the support of a land owning class. In those areas where the bulk of the farmers are tenants, the success of the projects are ultimately dependent on the Government's will to reform and its capability for reform.

Activities

The activities through which the goals of the minimum package projects are to be reached, are laid down in the Master Plan and in the Plan of Operation for EPID. In a 'model' project area, the following activities and personnel will be found:[51] The main activities include i) the demonstration of the effects of using fertilizer in cereal production, ii) the provision of inputs and a credit programme to make it possible for the peasants to benefit from the innovations and iii) a co-operative promotion programme. Every minimum package project will, as a rule, cover an area with 10,000 peasant families. The project personnel consists of one supervisor, five extension agents and up to five marketing assistants. The extension agents run the demonstration activities. They cultivate a demonstration field where the effects of fertilizing are shown. They also co-operate with and advise the model farmers (chosen by the project at the proposal of the local peasants), who are the ultimate link between the project and the population. The commercial activities (distribution of inputs and credit operations) are run by the marketing assistants. They arrange credit agreements, transmit supplies (fertilizer) to the peasants and collect repayments. The project area is divided into five extension areas/marketing centres. In each of these one extension agent and one marketing assistant are responsible for the contacts to five selected model farmers and to the approximately 2,000 peasant families.

The main activities are closely connected. Their aim is to increase and improve the production of cereal crops cultivated by the peasants and to make the producers dependent on an emerging money economy and a market system.

The inputs, which the project supplies, consist mainly of fertilizer but also of seeds and farm implements. The credit programme is concentrated on these inputs. From the project, farmers can get credits in kind while the repayments are collected in cash. The repayment period is one year and the interest rate is 12 %.[52]

The credit regulations require that each applicant must pay in cash 25 % of the value of the inputs before the fertilizer bags are handed over to him.[53]

The credit regulations are influenced by the experiences of CADU. A person who owns and cultivates more than 20 hectares is not qualified to apply for credits from the minimum package projects. Each borrower must have two personal guarantors. If the borrower is a tenant he must have a written lease agreement with his landlord (model leases have been worked out by EPID). One of the tenant's guarantors must be the landlord or the landlord's local representative. A landlord who is competent to apply for a loan (i.e. he cultivates himself less than 20 hectares) must sign acceptable written lease agreements with all his tenants within the project area before he can get a loan.[54]

These regulations are intended to i) prevent large landowners from dominating the credit programme, ii) force those large landowners who are qualified to apply for loans (those personally cultivating less than 20 hectares) to emancipate their tenants and iii) guard the financial operations of the projects (it is obvious that a tenant without written lease agreement is an unsafe investment object—he can be evicted after having spread the fertilizer on his fields). Point iii) turns our interest to the financial aspects of the credit programme. EPID co-operates with the AID Bank concerning the purchase of fertilizer on the international market. The Bank and EPID have signed an agreement according to which the former has appointed the latter as its agent for the provision of credit to farmers in the project areas.[55] The agreement stipulates that unless at least 94 % of the value of the credits distributed to farmers in a minimum package project area is repaid within due time (12 months + 2 months) the credit operations shall be discontinued. The same conditions are valid for the sub-divisions of a minimum package project. If 95 % of the amount due in one marketing centre area is not repaid within the agreed time, no further credit operations shall be undertaken there.[56]

One reason for these conditions is said to be the encouragement of "the principle of mutual responsibility between farmers for their loans".[57] Another (not mentioned) reason is that EPID and the AID Bank are anxious about their money.

The regulations concerning downpayment, written lease agreements for tenants and the principle that the inability of a few borrowers to repay their loans shall have negative consequences for all borrowers in the area, are assuredly justified from a narrow banking viewpoint. Nevertheless, the two first mentioned of the regulations create an inner contradiction in the project activities. On one hand, it is said that the activities shall benefit the farmers in the low income brackets. On the other hand, these very regulations can prevent farmers in the target group from participating in the credit programme.

The projects also sell fertilizer for cash. Any farmer can buy as much fertilizer as he whishes if he can pay for it and if there is enough fertilizer in the project store.[58]

Co-operative Promotion

In addition to the above mentioned activities, the minimum package projects are to assist in the creation and management of co-operative societies. These vehicles for the participation of the local population shall in theory have the ultimate function of making the projects superfluous—by taking over their activities. According to the plans co-operative societies are to take over the projects' role as a link between the AID Bank and the small farmers.

> "It is the intention of the Bank, as soon as co-operatives can be formed in the various project regions . . . to cease to use the Department [EPID] as its agent and to deal directly with the co-operatives".[59]

The co-operatives, when formed, shall be responsible for the following activities: applying for funds from the AID Bank, approving farmers' (= co-operative members') credit applications, accepting downpayment from farmers, distributing supplies to farmers, buying grain from farmers, collecting re-payments from farmers and distributing profits to farmers after debt deductions.[60]

These activities show that the co-operatives are supposed to have only commercial functions. The co-operative societies are to be formed during the first three years of the project. At the end of the third year the co-operatives are expected to take over the role of the project.[61]

To form co-operative societies is, of course, very desirable. But success, from the peasants' point of view, is dependent on what kind of power structure will be established within the co-operatives. The target group of the projects comprises illiterate peasants, ignorant of modern organizations and used to a low position in their hierarchically structured rural society. If the co-operatives are formed as a result of persuasion by the project personnel and local authorities, without preceeding political education of their prospective members, they run the risk of not being genuinely popular. It is not incongruous to assume that local 'big men' will establish themselves as top functionaries in the co-operatives. Imagine a landlord as the president of the association and his tenants as the ordinary members!

Without complementary reforms, the establishment of co-operative societies may mean only that the age-old exploitation of the poor peasantry will continue—adapted to modern organizational techniques.

Next we turn to a detailed study of two minimum package project areas. The areas are Bako and Shashamane. I selected them for study in 1971 from among the then existing nine project areas.[62] The analyses will illustrate the problems that face the minimum package approach when it is tried in a stratified society where the target group of the activities hold a subject position. Concerning the generality of the studies, I believe that the experiences from Bako and Shashamane are relevant for most peasant regions in Ethiopia, where land is privately owned and landownership is concentrated.

Notes to chapter 7

[1] *TFYDP*, p. 373.

[2] Bengt Nekby, *CADU, An Ethiopian Experiment in Developing Peasant Farming* (1971), p. 14.

[3] *TFYDP*, pp. 40 f., 371 ff.

[4] WADU was established in 1970. The project is jointly financed by the Imperial Government and the World Bank Group. Its activities include agricultural extension and a settlement scheme in Wolamo *awradja*, Sidamo province. WADU will not be analyzed here.

[5] *CADU Project Description* (April 1971).

[6] EPID, *Annual Report for 1971/72 Financial Year* (1973); EPID, *Annual Report for 1972/73 Financial Year* (1973); *Work Programme and Budget for EPID 1973/74 and 1974/75* (1973).

[7] The presentation of CADU's activities is based on the following sources: *CADU Project Description* (October 1971); *CADU Preliminary Final Report for the Period 1967–70* (1970); *CADU–a Brief Description* (1973); Nekby, *op.cit.*

[8] *Kungl. Maj:ts proposition nr. 100 av år 1962. Kungl. Maj:ts proposition till riksdagen angående svenskt utvecklingsbistånd, given Stockholms slott den 23 februari 1962.*

[9] Translation by the author from *Ibid.*, p. 7. The original Swedish version reads: "Det är vanskligt att ange sociala och politiska mål för biståndsgivningen. Det är inte givet att de sociala och politiska system och principer till vilka vi ansluter oss är för alla u-länder ändamålsenliga eller uppnåeliga. Trots det kan det vara rimligt att söka inrikta biståndsgivningen så att den, så långt man kan bedöma, medverkar till en samhällsutveckling i politiskt demokratisk och socialt utjämnande riktning. Det är inte i överensstämmelse vare sig med det svenska biståndets motiv eller med dess mål att det bidrager till att konservera en framstegsfientlig samhällsstruktur."

[10] *Report No. 1 on the Establishment of a Regional Development Project in Ethiopia* (1966), Part II, p. 189.

[11] *Ibid.*

[12] *CADU, Plan of Operation for Regional Agricultural Development in the Chilalo Awraja, Arussi Province, Ethiopia 1967–70. Annex to the Agreement of September 8, 1967 between the Swedish Government and the Imperial Ethiopian Government on cooperation in the field of regional agricultural development* (1967), p. 1.

[13] The Swedish-Ethiopian negotiations which eventually led to the establishment of CADU and the influence exercised over the formulation of goals and activities by the involved parties open up interesting fields for research, which I cannot touch upon here.

[14] Henock Kifle (1971), *op. cit.*, Appendix II.

[15] *Ibid.*, Appendix III.

[16] Göran Bergman and Håkan Lindqvist, *Credit Situation in Chilalo Awraja* (1969), p. 39.

[17] *Survey of Arussi*, pp. 18, 20.

[18] Nekby, *op. cit.*, p. 60.

[19] *CADU Annual Reports 1971/72 & 1972/73* (1973), p. iv.

[20] Information obtained from EPID's headquarters.

[21] *Final Report on the Appraisal of CADU and EPID* (1974), p. 8.

[22] Henock Kifle (1971), *op. cit.*, p. 15.

[23] *Ibid.*, p. 41.

[24] *CADU General Agricultural Survey 1972* (1973), pp. 7 f.

[25] This conclusion was drawn by CADU's Evaluation Section already in 1971. See Henock Kifle (1971), *op. cit.*, p. 41.

[26] *Report No. 1 on the Establishment. . .*, Part II, p. 389; *CADU Plan of Operation 1967–1970*, p. 10.

[27] The gradual playing down of the Awradja Development Committee is illustrated in Peter Wallensteen–Bo Schiller, *Fallet CADU, Organisation, Produktion, Folkligt Deltagande och Ägarstruktur i ett Regionalt Landsbygdsutvecklingsprojekt* (1972), Del III, pp. 10 f., 44–51.

[28] *Tentative CADU Programme 1970–1975* (1969), p. 12.

[29] *Preliminary Report of the Appraisal Team on CADU* (1969), p. 3.

[30] *CADU Annual Reports 1971/72 & 1972/73,* p. 47.

[31] *Ibid.*

[32] *Final Report on the Appraisal of CADU and EPID,* p. 29.

[33] The project manager writes about the *awradja* and *woreda* officials: "Collaboration with the local administration has not proceeded without friction. On account of its bureaucratic, authoritarian structure and sometimes corrupt character, the local administration is often mistrusted by the farmers, and close collaboration would have implied certain risks to the project. Attempts to inform the local administration about the activities of the project and to interest them in it have met with a rather cool reception". Nekby, *op. cit.*, pp. 83 ff. Another indication to the reluctance of the Imperial Government to the goal of popular participation is that the Ministry of National Community Development and Social Affairs has refused to properly register two of the pre co-operatives in Chilalo. See *CADU Annual Reports 1971/72 & 1972/73,* p. 48.

[34] An indication of this is that the Swedish project management decided not to use the traditional peasant voluntary organizations in Chilalo as basis for co-operative societies, although these organizations were well functioning and furthermore well documented. The anthropologist Arne Lexander has described them in detail in his study *The Changing Rural Society in Arussi Land,* which is a CADU publication.

Professor Milosavljevic, one of the members of the evaluation team on CADU 1967–70, made the following statement on CADU's influence on popular participation: "Certainly a production-oriented project such as CADU, under what might be described as the paternalistic control of agricultural technicians, cannot and has not contributed to improving this ability [popular participation]". See *Preliminary Report of the Appraisal Team on CADU,* p. 30.

[35] The number of registered tractors in Chilalo did not exceed 35 until the year 1968/69 when it suddenly jumped up to 61—an increase of almost 100 % in one year. During the following years the rapid increase continued. These figures have been compiled from the Economic Research Division within the Ministry of Agriculture and are quoted in Haile Wolde, *op. cit.*, pp. 15 f.

[36] *CADU General Agricultural Survey 1972,* p. 25.

[37] *Ibid.*, pp. 25 f.

[38] Levon Debelian, *The Impact of Agricultural Mechanization on the Traditional Society in Ethiopia* (1973), p. 18.

[39] Henock Kifle, *Investigations on Mechanized Farming and its Effects on Peasant Agriculture* (1972).

[40] *CADU General Agricultural Survey 1972,* p. 23.

[41] *Ibid.*, p. 24.

[42] Meheretab Tekie, *A Case Study of a Small-Scale Commercial Farm* (1969).

[43] *Plan of Operation for EPID* (1971), p. 1.

[44] *A Master Plan for the Extension and Project Implementation Department of the Ministry of Agriculture* (1970), p. 4. (Hereafter: *A Master Plan for EPID*).

[45] *Ibid.*, p. 7.

[46] *Plan of Operation for EPID,* pp. 7 f.

[47] *A Master Plan for EPID,* p. 6.

[48] *Plan of Operation for EPID,* p. 8.

[49] *Ibid.*

[50] *Ibid.*, p. 7.

[51] *A Master Plan for EPID,* pp. 10 f.

[52] *Plan of Operation for EPID,* Appendix 1 B, pp. 2 f. The interest rate is 1 % per

month, but at least 8 % is charged however early the loan is repaid.

[53] EPID, *Manual for the Distribution of Inputs* (1973), p. 5. (Hereafter: EPID, *Manual...*).

[54] *Ibid.*, pp. 3 ff.; *Plan of Operation for EPID,* Appendix 1 B, pp. 2 f.

[55] *Plan of Operation for EPID,* Appendix 1 B, p. 1.

[56] *Ibid.*, p. 4.

[57] *Ibid.*, p. 3.

[58] *EPID, Manual...*, p. 5.

[59] *Plan of Operation for EPID,* Appendix 1 A, p. 5.

[60] *A Master Plan for EPID,* p. 21.

[61] *Ibid.*, p. 11.

[62] Of the nine minimum package projects which were established in 1971 three were located in communally owned areas (Enda Selassie, Fenote Selam and Gondar) while six were located in areas where private ownership prevails (Asendabo, Bako, Haik, Kersa, Shashamane and Tulu Bollo). I selected Bako to illustrate a feudal structure and Shashamane to illustrate a structure where feudalism has been penetrated by large scale commercial farming.

Chapter 8 Bako Minimum Package Project

8.1. An Outline of the Social Structure

General Information

Bako minimum package project started its activities in the middle of 1971. The project centre is situated in the town of Bako, 200 km west of Addis Ababa on the main road Addis Ababa–Ghimbi. The administrative boundary between Shoa province and Wollega province runs immediately west of the town. In spite of several administrative divisions splitting up the area, Bako is part of the greater natural and cultural geographical unity which is constituted by the western part of the Shoan plateau. This plateau, approximately 2,000 meters above the sea, is cut through by numerous valleys. On the higher altitudes the climate is temperate and cereals are the main cultivated crops. In the valleys the climate is hot and allows the cultivation of tropical crops. The rainy season lasts from May to late September.

The bulk of the population belongs to different groupings within the Oromo family of tribes and Oromo is the language of the population. The official, administrative language is–here as everywhere else in the empire–Amharic, and Amharas are found among Government officials at all administrative levels, as well as among merchants and wealthy landlords.* The majority of the population belong to the Ethiopian Orthodox Church but also Islam has many followers in some localities.

History

Since the sixteenth century the western Shoan plateau has been the home of Oromo tribes. These were first almost exclusively pastoral but shifted gradually the emphasis of their economy to agriculture. In the middle of the nineteenth century they were fully settled and political authority among them became increasingly dependent on ownership of land. Chieftaincies and petty kingdoms developed. In the late nineteenth century the region was incorporated into Menilek's expanding empire. The incorporation needed little violence. Menilek's general Gobana, himself of Oromo stock,

* Amhara peasants are also, in moderate numbers, migrating from Godjam across the Nile into northern Wollega in search for new lands to cultivate.

succeeded in persuading the Oromo chiefs to submit to Menilek's authority in exchange for Imperial privileges.

After the incorporation the land was redistributed. It is claimed that 3/4 of the land was reserved for the Crown while the remaining fourth was divided between Oromo chiefs, who were now assimilated into the Imperial administration.

Parts of the Crown land were kept by the Emperor personally, and other parts were given to members of the Imperial family, highranking officers and other persons of influence. These in their turn allotted pieces of land to their subordinates. At the bottom of this hierarchy were the Oromo peasants, now converted to *gabars*.

Haile Selassie's land grant orders in the 1940's and 1950's secured the ownership of these lands for the descendants of the original grantees. The *gabar* system was abolished and some peasants managed to get ownership rights to their holdings while the majority became tenants.

Land Tenure Pattern in the 1960's

Since the project operates within two *awradjas* (Jibat & Mecha in Shoa province and Nakamte in Wollega province) some statistical data from these *awradjas* will be presented as an introduction. The data emanate from sample surveys carried out by the Central Statistical Office in 1963–64. The figures should be considered as rough indicators.

According to the surveys less than 20 % of the total area of the two *awradjas* were under cultivation in the mid 1960's.[1] The peasant holdings were small; 95 % of the agricultural holdings in Jibat & Mecha comprised less than five hectares while no peasant holding exceeding four hectares were reported in Nakamte.[2] Cereal crops dominated agricultural production, teff and barley in high altitude areas and maize in the valleys.[3] The land tenure pattern is indicated in the following table.

The total number of holdings in Jibat & Mecha and Nakamte was reported to be 76,000 and 37,000 respectively.[4]

The large proportion of tenant operated holdings indicates the oc-

TABLE 6. Percentage distribution of peasant holdings by tenure in Jibat & Mecha and Nakamte *awradjas*.

Awradja	% of holdings by tenure owned	 rented	 partly owned/ partly rented
Jibat & Mecha	32 %	56 %	12 %
Nakamte	27 %	69 %	4 %

Source: Survey of Shoa, p. 29; Survey of Wollega, p. 29.

Map 8. PLACES REFERRED TO IN THE TEXT

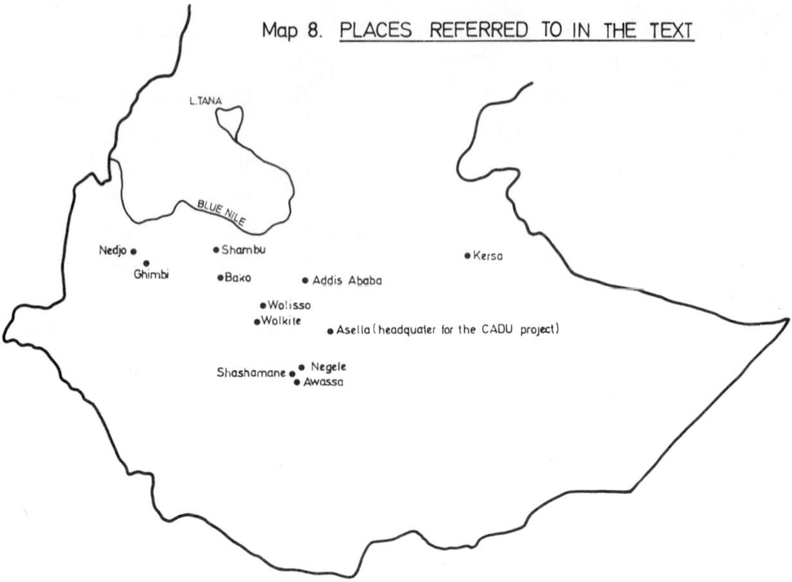

currence of absentee landlords. It was reported that absentee landlords owned 52 % of the measured area in Nakamte.[5] The largest area in Nakamte owned by one individual exceeded 3,000 hectares.[6]

The majority of the tenants paid rent by share-cropping. Most of them kept bullocks and farm implements of their own. The rent was then, in most cases, less than 50 % of the harvest. In Jibat & Mecha many tenants paid between 10 % and 25 % of the harvest.[7] With the gradual coming of the money economy a form of rent payment called 'contract' has gained in importance. In this form rent is paid as a fixed sum of money. Compensation for improvements made by the tenants was as a rule not paid by the landlords. According to a survey of Nakamte, compensation was paid to 11 % of the surveyed tenant holdings.[8]

Two socio-economic surveys carried out in 1969 and 1971 give the following data on the immediate rural surroundings of Bako. The 1969 survey covers 118 peasant families out of which 70 % were tenants and 30 % owner-cultivators.[9] None of the families cultivated more than 2.6 hectares (teff, maize and noug) although some owner-cultivators owned a larger area. The greater part of the production was used for subsistence while some was sold on the market in Bako to cover taxes and the purchase of necessities as salt, tobacco and coffee. The tenants used part of their production for paying rent, 72 % of their lease agreements were oral and 28 % written, 88 % of the tenants paid rent in the 'contract' form, i.e. each year the landlord and tenant agreed on a fixed sum of money or amount of grain to be paid at the end of the agricultural season. The frequency of evictions was not high, 80 % of the tenants had been residing on the same holding for more than 4 years and 68 % for more than 11 years.

The 1971 socio-economic study covers 196 families along the road Bako-Jere,[10] (Jere is situated 11 km north of Bako). Out of these families 63 % were Oromos and 25 % Amharas, 81 % belonged to the Ethiopian Orthodox Church and 17 % to Islam. A remarkably high proportion (40 % of the heads of the households) were not full time farmers; they earned their living from trade and handicraft. In 39 % of the households there were income earners other than the head of household. Job opportunities were provided by two large scale mechanized farms: those of the Evangelical Church Mekane Yesus and the Ethio-German agricultural station. These hire more than 300 workers on a temporary basis during the agricultural season. When it comes to the fulltime farmers 72 % of them were tenants. None cultivated more than 3 hectares. Only a few owner-cultivators had a cultivated area exceeding 4 hectares. Though the holdings were small they contained also idle land. It was estimated that idle land represented between 20 % and 30 % of the holdings.* In the survey three reasons are put forward to explain this paradox: 1) shortage of manpower because the two mechanized farms attract labour during harvest time, 2) shortage and weakness of work oxen during the dry season when the ploughing has to be done, 3) incapacity of most of the farmers to raise investment capital due to their status as tenants. The tenancy agreements as reported in the study do not differ qualitatively from those described in the 1969 study; 40 % of the tenants paid a fixed sum in cash ('contract') while the rest were share-croppers. The most common share-cropping agreement was that the tenant handed over one half of his harvest to the landlord (*ekul arash*). In addition the landlords received tithe, i.e. one tenth of the total yield from the tenant's holdings.

The study reports that 60 % of the tenants in the sample had lived on their present holdings for more than 4 years and 34 % had lived on their holdings for more than 11 years. The study comments that "this suggests a rather slow turnover of tenants or a low incidence of actual eviction".[11] As will be shown below, this comment had lost its validity one year later.

* The word 'idle' which is used in the report is misleading. It must be assumed that much of the 'idle' land is used for grazing. The peasants keep cattle, sheep, goats, mules and donkeys. Crop growing is complemented by animal husbandry in the peasant economy.

The farmers in the study grew teff, pepper and noug for the market in the town of Bako and sorghum and maize partly for their own subsistence and partly for the market. Due to lack of storage capacity they had to sell a part of their maize just after harvest, i.e. when the price is at its lowest. Later on in the year they ran out of maize for their own consumption and had to buy maize from merchants at a higher prize.*

The two mechanized farms which have existed in the area for a number of years do not seem to have spread technical innovations to the peasants. On the contrary, the 1971 study reports that most of the interviewed farmers said they had never received instructions in the use of fertilizer.[12]

An attempt will now be made to analyze the power structure which is indicated by the data presented above. The point of departure is the large number of tenants, which is an undisputable fact.[13] This expresses indirectly that there are people who live on incomes from leased land. The official surveys do not contain detailed information on large landlords, it is even difficult to obtain accurate figures on their number and on size of owned land. I have, however, gathered information about the landlords in Bako area from other sources. The information obtained forms a pattern, the profile of which will now be sketched.[14]

By far the largest landlord is the Imperial family. In and around the project area in Sheboka (close to Bako) appr. 12,000 hectares are registered in the name of the late Prince Sahle Selassie. The holding is administered by a representative on behalf of the heirs. The tenants cultivating the land of the Imperial family pay rent, agricultural income tax and tithe, and they cultivate fields for the personal use of the administrator. Even if the tenants knew that some of these obligations have been declared illegal by Parliament (personal labour and tithe) they have nowhere to take their complaints – it is the highest authority in Ethiopia which has sanctioned their obligations. Indeed, the representative even keeps a private prison where he temporarily locks up trouble makers.

Second in size after the Imperial land come the lands of the *balabbats*. In Bako their function is to collect taxes from the peasants for the Ministry of Finance and to communicate Government decrees and orders to the

* A similar pattern is reported from Gidami *woreda* in western Wollega. One of the villages there is owned by one single landlord. The tenants are share-croppers, they pay 50 % of their harvest (maize) as rent. When the landlord has taken his share the tenants can subsist from 4 to 5 months on the remainder. When the food is on the decline the tenants walk to Gidami town to get occasional jobs. They are paid from Eth. $ 0.30 to 0.50 a day. Their income is used for buying maize at the market in Gidami at a price of Eth. $ 10–15 per quintal (*Report on a Survey to Begi and Gidami woreda*). The tenants are exploited both as producers and as workers. The maize they buy at the market may be the same maize as they have given away to the landlord as rent.

peasants.[15] Within the *woredas* where Bako minimum package project operates there were in 1973 a total of 29 *balabbats*, each of which had an average area of 102 *gashas* under his administrative responsibility.[16] Among the *balabbats* there were both Oromos and Amharas.

At the local level the *balabbat* is an undisputable authority for the peasantry and he can operate quite independently of the official local Government. In some places the word of the *balabbat* carries more authority than that of the *woreda* governor and the former can put pressure on the latter. In the Bako area, most *balabbats* use their authority to exploit the peasants. They can for example require all peasants, owner-cultivators as well as tenants, to work for them on certain days. The *balabbats* in Ejaji and Jere (where the project has marketing centres) have private prisons where they use to keep peasants jailed who question their authority.[17]

The *balabbats* are at the same time large landlords. In the vicinity of Bako the largest landowner is a *balabbat* who owns 33 *gashas*.[18] Tenants on the *balabbat's* land have to deliver the produce they pay as rent to the *balabbat's* store and they have to perform all agricultural work on the fields which are used for the needs of the *balabbat's* own household.

All landlords are of course not *balabbats*. There are a number of other persons who own large areas which are leased out to tenants. As to historical origins, the local landlord class around Bako can be divided into two subgroups, i) the descendants of Oromo chiefs and Shoan military officers who received land during Menilek's time, ii) individuals, not necessarily from Bako, who in more recent times have obtained grants from Haile Selassie. In both cases ownership has been given on political grounds without consideration of the peasants who actually live on the land.

My informants emphasize the precarious social and economic position of the tenants. They seldom have any alternative but to submit to the demands of their landlords. When disobedience or protest occurs the landlord (or his local representative) can react by terminating the tenancy relationship. The conditions of tenancy signify an exploitation of which the tenants are not unconscious, but the termination of the tenancy relationship would mean an acute catastrophe for the tenant family. Agriculture and animal husbandry are the only means of subsistence available to them. For an evicted tenant family there is only one way out—to look for a new landlord and ask for a piece of land to cultivate. If an agreement materializes, it is easy to understand that the parties are not equals. The landlord has the power in his hand, and this is reflected in the tenancy conditions.

This means that evicion is an ever present latent threat, which makes the tenants obedient to their obligations.

In order to obtain a more vivid picture of the social structure in the area around Bako, findings from three cultural geographical studies of villages not far from Bako will be summarized below. The studies have been carried out by Ethiopian students at the turn of the decade. Each of the studies covers a continuous area occupied by ten peasant families.

Three Villages in Jibat & Mecha

The first village is Lege-Batu situated 84 km west of Addis Ababa on the main road to Ghimbi[19]. The land around this village was in the early twentieth century Imperial domain. The peasant's production was used to supply the Imperial palace in Addis Ababa with food stuffs. Later on the Empress Menen received it as her personal property. When she died her children—the princes Makonnen and Sahle Selassie and the princess Tenagne Werk—inherited it. They have in turn granted some part of the land to servants while their children have inherited other parts.

A study of ten peasant families living in a continuous area comprising 91 hectares was made in 1970. All ten farmers were tenants cultivating the land of three absentee owners.

The tenants had until recent times been share-croppers, giving one-third of their harvest to the local representative of the owner as rent. Some years before the study was made the owners had abandoned the share-cropping system in favour of payment of a fixed amount of grain or money. This recently introduced system was called 'contract' among the farmers. According to the author of the study, the fixed amount of grain to be paid was in 1971 a somewhat better deal for the farmers than 1/3 of the harvest would have been. But the owners are free to increase the amount to be paid.

The three owners are not personally present in Lege Batu. Their interests are handled by a local representative, who collects the rent. According to the representative a tenant may stay in the area as a farmer as long as he is hard working and pays the rent on time. If he is careless and not obedient enough, he can be evicted.

In addition to the rent, the tenants pay an agricultural income tax to the Government. This tax varies between Eth. $ 1.50 and 6. The owners pay tax too. This is, however, a negligible amount compared to their incomes from the leased land. The author of the study mentions that one owner in the neighbourhood paid Eth. $ 90 in taxes while he the same year collected a total of 45 quintals grain as rent. This amount was equivalent to about Eth. $ 1,000 at prevailing market prices.

The second village is Botche[20], a few kilometers east of Lege Batu. The ten peasant families in the study are all tenants and cultivate the land of one single owner. The owner has received the area as freehold from princess Tenagne Werk. He lives in Addis Ababa. Earlier the tenants were share-croppers, they paid one-third of their harvest as rent. In 1970 the owner introduced the contract-system, so the rent is now a fixed amount of grain. The change from share-cropping to fixed rent was motivated by the owner with the argument that as long as share-cropping existed the tenants could afford to be lazy and produce little. The fixed rent is made so high that it makes them produce more in order to have something remaining for themselves.

The ten tenant families had gone together and rented the whole area paying 60 quintals of grain to the landlord (1971). The tenants decided at a meeting how much each of them shall contribute to the total amount. The lease was drawn up for one year and renewed annually. According to the author of the study the tenants did not like the new system of paying rent since they considered the new rent too high. But they accepted the system because they were afraid of loosing their holdings. They had decided to ask the landlord to reconsider his decision, but should he refuse they would pay if they could. If they could not, they would leave the area.

The local representative of the owner is a farmer on the land of his employer. His duty is to collect the crops given as rent. In return he pays no rent himself. The representative is Amhara, while the tenants are Oromos.

The third village—Facha—is situated about 100 kilometers west of Addis Ababa along the Nakamte road[21]. The ten peasant families surveyed live in an area comprising 140 hectares of which 48 are cultivated. The ownership structure in Facha differs from that of Lege Batu and Botche. The majority of the farmers are tenants, while a minority are owner-cultivators who own a few more hectares than they themselves can cultivate, and they therefore lease out their additional land. Some of the tenants cultivate land belonging to absentee owners who live in a town not far from Facha.

In Facha we find a situation where owners and tenants live and work side by side. The owners are also peasants even if they are better off than the tenants. In some cases owners and tenants are relatives. They are all Oromos.

The Facha tenants are share-croppers, most of them pay one-fourth of the harvest as rent provided that they use bullocks and farm implements of their own (*irbo arash*).

The study gives the impression that the relations between owners and tenants in Facha are not characterized by antagonism. The owners participate themselves in the daily toils of agricultural work and this makes them sympathetic to the problems of the tenants. Owners and tenants co-operate in the choice of crops for each agricultural season. Evictions occur only rarely. When an eviction takes place, it is only after the elders of the village have found that the tenant in question is misusing the farming plot.

The author of the study emphasizes that the decisive difference between owners and tenants in Facha is not economic but social. The owners are well respected, even if they are owners only of small farming plots. They are called upon to settle disputes, they are the informal authorities of the local society.

8.2. The Project's Field Activities

Coverage

During its first year of activity Bako minimum package project established extension areas/marketing centres in five places: Bako, Jere, Anno, Ejaji and Sheboka. In 1972 the project activities expanded to Sire and in 1973 to Kejjo. Thus, in late 1973 the project activities covered an area 10 km x 71 km along the Addis Ababa–Ghimbi road and an area 10 km x 18 km along the Bako-Jere road.[22] An agricultural extension agent and a marketing assistant were stationed in each of the extension areas/marketing centres. The supervisor resided in the town of Bako. Bako had approximately 4,000 inhabitants in 1972.[23]

The Credit Programme

The project's main activity has been demonstration of the effects of fertilizer (on the project's demonstration fields and the model farmers' fields) and provision of fertilizer to the farmers on credit and on a cash basis. Table 7 shows how the fertilizer has been distributed to different categories of farmers during the three first credit seasons.

Each extension area is estimated to cover about 2,000 peasant households, i.e. the total number of participants should be compared with the estimated total number of peasant families within the project area which was 10,000 in 1971, 12,000 in 1972 and 14,000 in 1973.

Three facts stand out from the table: the total number of participants is small, the rate of increase over the years is slow and the tenants are virtually absent in the credit programme.

The expansion of the credit activities from 1971 to 1973 is explained by the increased number of extension areas. The increase in 1973 is to a great extent explained by the inclusion of 49 farmers from the Evangelical

TABLE 7. Bako minimum package project fertilizer distribution 1971, 1972, and 1973. No. of participants.

Exten-sion area market-ing centre	1971			1972			1973		
	cash	credit		cash	credit		cash	credit	
		owner/ culti-vators	tenants		owner/ culti-vators	tenants		owner/ culti-vators	tenants
Bako	2	13	–	2	14	–	13	19	–
Anno	–	6	3	4	19	–	4	92	–
Jere	–	11	1	–	13	–	3	13	–
Ejaji	–	31	3	1	25	–	1	23	–
Sheboka	–	10	2	14	22	–	6	16	–
Sire	–	–	–	3	12	–	2	25	–
Kejjo	–	–	–	–	–	–	–	6	–
Total	2	71	9	24	105	–	29	194	–

Source: The project centre in Bako[24]

Church Mekane Yesus' settlement in Ano-Gambela[25]. In the original five extension areas the increase has been negligible, in one area the participants have actually decreased in number.

Before probing into the reasons for this blockage, the actual participants will be examined. The table classifies 9/10 of the credit takers in 1971 as owner-cultivators. One-third of these owner-cultivators cultivated more than ten hectares, thus placing them in the small stratum of rich peasants. In 1972 the proportion of rich peasants among the credit takers was about the same, while their proportion decreased in 1973[26]. The cash buyers have increased considerably, from a modest beginning in 1971 to 29 in 1973. Peasants who can afford to buy fertilizer cash are by definition rich in the Ethiopian context. In particular the cash buyers in Sheboka extension area were wealthy people[27]. Nine credit takers were registered as tenants in 1971. In the following years all credit takers claimed to be owners, but the project staff believed that a few tenants hid among them although it has been impossible to check this[28]. Nevertheless, the fact remains that Bako minimum package project has not reached its target group: In 1973 the project covered some 14,000 peasant households out of which approximately 70 % or 9,800 were tenant families. Instead of reaching these, the project has given service to a small number of landowning peasants, among whom a considerable part belonged to the relatively wealthy stratum.

In order to gain an understanding as to why the Bako minimum package project has been a failure (at least up to 1973) it is necessary to focus on the social structure of the agricultural society around Bako and on the credit regulations of the minimum package project.

One barrier blocking the way to participation in the credit programme is the regulations concerning down-payment. A credit taker must pay in cash 25 % of the value of the credit before the fertilizer bags are delivered to him. This regulation excludes the poorest peasants, those living in the utmost periphery of the monied economy. The fertilizer is distributed in bags containing half a quintal (= 50 kg). EPID's price for one bag was Eth. $ 19* during the period 1971–73. In order to yield good results one hectare needs two bags (one quintal) of fertilizer, the amount recommended by the extension agents. If a peasant wants to take credit for the minimum amount of one quintal he had to make a downpayment of Eth. $ 9,5.

The Gross National Product per capita in Ethiopia was stated to be Eth. $ 165 in 1969.[29] A case study from western Shoa (a village close to Ambo, approximately 100 km from Bako) classifies the interviewed peasants in the following income groups:

TABLE 8. Income groups in Mankata.

Income group Eth. $ year	Households number	%	Total income Eth. $	%	Assumed average income within group
– 100	25	22	1,500	5	60
101–200	29	26	4,350	15	150
201–300	28	25	6,720	22	240
301–400	12	10	3,960	13	330
401–500	8	7	3,400	11	425
501–	11	10	11,000	34	1,000
Total	113	100	30,930	100	270

Source: Asfaw Edessa, *A Micro-Geographical Study of Mankata* (1970).
The table has been elaborated by Lars Bondestam.

Forty-eight percent of the households had an income below Eth. $ 200. The poorest group of peasants, those earning below Eth. $ 100 comprised 22 % but had only 5 % of the total income at its disposal. For peasants in this group the downpayment is equivalent to one month's income. It is beyond doubt that this group cannot afford to participate in the credit programme and it is dubious whether those earning between Eth. $ 101 and 200 can afford to participate.

The regulation concerning downpayment strikes at both owner-cultivators and tenants. For the latter group, another condition for participation in the credit programme is required. Tenants must be able to pre-

* This was the price for D.A.P. which is the most generally used fertilizer in Ethiopia.

sent a written lease agreement on their holding, signed by the landlord or his local representative. The landlord must also be one of the guarantors of the borrower.

Many large landlords in Bako are negative to the concept of written lease agreements since this would increase the security of the tenants. This holds true for the representative of the Imperial land in Sheboka as well. It has been reported that the large landlords consider the activities of the minimum package project aimed against their own interests. The project refuses to give them credit and concentrates their activities on the small farmers. This is considered hostile. It is also not unusual that the large landlords see the project activities in a longer perspective. The project aims at persuading the landlords to give their tenants written lease agreements. From the point of view of the landlords, this would be equal to retreat from their present power position. It would result in an increased security in land for tenants and make them inclined to demand further rights. In the extension of this perspective a change in the whole ownership structure of rural Ethiopia emerges—the tenants would at last demand ownership rights to their cultivated fields and redistribution of holdings. Such a vision is a nightmare for the large landlords. Those of them who look at things this way consider the minimum package projects not only credit institutions for small farmers but also, in the long run, instruments of political subversion.

Co-operative Promotion

For the philosophy of the package approach to agricultural development the co-operative idea is central. In Bako, activities aimed at formation of agricultural co-operative societies did not start until late 1972[30]. The machinery which the project had at its disposal for this purpose, consisted of the supervisor, a cooperative organizer and the marketing assistant in the marketing centres. During the period between harvests in 1972 and 1973 they arranged meetings and gave lectures on the benefits of co-operatives. An average of one meeting a week was held in 5 of the 7 extension areas, the average number of peasants participating in the meetings was 30—40, the highest attendance being 105 peasants.[31] At the end of 1973 the meetings had not yet resulted in the formation of co-operatives or the election of executive committees[32]. The *balabbats* were reported to have discouraged peasants from taking part in the meetings, but no actual attempt at disturbing a meeting had occurred.[33] The failure to establish co-operatives must partly be due to the negative attitude of the local powerholders. For the peasants, to challenge the authority of landlords means to risk one's personal security.

The reasons for the modest progress in the field of co-operative promotion must also be looked for in EPID's policy. Until 1973 the promotion of co-operatives was more or less up to the particular minimum package personnel. Central EPID did not give detailed advice or education on how the project personnel should proceed in promoting co-operatives.*

8.3. Mechanized Farms

When I visited Bako in December 1971 the feudal structure was still intact. Low-yielding smallholdings operated by tenants were the rule. Large tracts of land were also used for grazing or lay idle covered with bush.

When I revisited Bako in May 1972 the situation had changed.[34] The first commercial agricultural entrepreneurs had discovered the area. Parts of the Imperial land in Sheboka had been contracted by four individuals. The total area to be affected was about 900 hectares. The transformation of this land into mechanized farms was to take place gradually. In May the preparations were underway; 107 tenant families (some 500 individuals) cultivating areas which had been contracted and selected for tractor cultivation for the next season, had been told by the entrepreneurs to leave their holdings since their leases had been terminated.

The entrepreneurs were not local people, they came from Addis Ababa. Two of them were young, educated men, while the other two (one being a retired colonel, the other a vice minister in the Government) were characterized as "experienced exploiters" by my informants. The background of the commercial farmers show that the inspiration to take up commercial, mechanized farming in this case did not emanate from the minimum package project but from the Imperial Government's policy of encouraging large scale commerical farming. The fact that the mechanization took place on Imperial land demonstrates that the highest authority in the country approves of this policy in spite of the evictions which are bound to accompany it.

When I again revisited Bako in November 1973 the mechanization had proceeded further[35]. In all, approximately 360 hectares were put under tractor cultivation (the commercial farmers had 2 to 3 *gashas* each).

In 1972 maize and sorghum were the principal crops grown, while haricot beans were introduced in 1973. The swing to this latter crop shows that the commercial farmers are responding to the latest market signals. In 1972 the price of maize had been exceptionally low, Eth. $ 5 per quintal at Bako market[36], while at the same time export possibilities for haricot beans were opened.

* For a further discussion on this point see chapter 10.

The coming of commerical farming has brought with it an increasing misery for the peasants who have been affected by it. The 107 families threatened by eviction in 1972 have been affected in two ways:[37] A minority has disappeared. These were tenants whose fields and homes had been converted to tractor ploughed fields. They had no other choice but to take their cattle and leave. The majority of the families lost their fields but were still around in the area. These were families whose homes were situated in places not suitable for tractor cultivation. The commercial farmers invited these families to stay and perform temporary jobs during weeding and harvesting time on the newly mechanized farms. As long as maize is cultivated by the commercial farmers quite a number of daily workers are needed seasonally. The cultivation of haricot beans can on the other hand be more mechanized and needs fewer daily workers. The temporary jobs offered to earlier tenants were of an unqualified nature (weeding, watching crops against monkeys etc.) and the payment was minimal. A limited number of qualified workers (tractor drivers) were brought in from outside the region.

For the agricultural year 1974 the commercial farmers are going to enlarge the mechanized area. An additional one hundred tenant families have been told that their lease agreements will be terminated by 1974.* Among them is the minimum package project's model farmer in Sheboka. The project is considering closing down its activities in Sheboka.

The coming of mechanized farming has also had negative consequences for peasants whose fields have not been directly affected. The peasant economy is dependent on animal husbandry as a complement to crop growing. The peasants' cattle are grazed on land belonging to landlords. In Sheboka, one of the commercial farms is so situated that it blocks the traditionally used road to grazing areas for a number of families. These peasants have been forbidden to let their cattle pass through the commercial farm. This means that the cattle are confined to a more limited grazing area which also belongs to the commercial farmer in question. In this particular area a monopoly situation concerning grazing land has arisen and the commercial farmer is in a position to demand increased rents for grazing.

* News reports in June 1974, immediately before the printing of this book, relate violent clashes between commercial farmers and tenants threatened by eviction in Sheboka. Employees of the commercial farmers set fire to tenants' huts and tenants reacted by hindering tractors from plowing. The police was brought in and many tenants were arrested. (Anders Johansson, Gobena överger aldrig sin jord, article in *Dagens Nyheter*, June 5th, 1974 and supplementary oral information.)

8.4. Conclusions

Bako minimum package project has—up to 1973—been a failure. The officially stated goals have not even come close to attainment. And the prospects for the future seem gloomy. The reasons have been looked for in the structure of the society where the project operates. This structure will now be summarized in theoretical terms.

The agricultural society in the region where the project operates is a class society. The productive class is the peasants, of which the overwhelming majority are tenants. A considerable proportion of the production of the tenants is appropriated by the class of medium to large landlords. The appropriation, which is exploitative, is not taking place by capitalistic means, but through a form of tribute.

The land is the private property of the landlords, the exploitative relationships penetrate from the landlords directly down to the individual tenant families. Because the land is owned by landlords, the village communities, within which the tenants are grouped, do not constitute entities of political significance that could protect the individual families.

A consequence of this particular structure is that the peasants show little initiative in agricultural production. The producers do not themselves benefit from an increase in their production.

Nor has the class of landlords tried to push the tenants into innovation for increased production. The landlords have confined themselves to developing the mechanisms of exploitation.

One example of this is the transition to the contract form of rent payment. The fixed amount of rent protects the landlord from a low rent in case of partial crop failures. And the landlords are free to increase the amount of rent which is contained in the contract. Another example is the interference of the landlords in the choice of crops to be grown on the tenants' holdings. From a village east of Bako it is reported that landlords force their tenants into monoculture. The tenants are forbidden to cultivate a variety of subsistence crops and are urged to grow cash crops only (teff and wheat).[38] This is because the landlords are not any longer interested in tributes in form of subsistence goods but only in crops which can be translated into money.

These aspects of exploitation, which seem to be of recent origin (the 1960's), illustrate something more than the landlords' greed. They are also peripheral reflexes of Ethiopia's increasing attachement to the international capitalist system. Also in rural areas the landlords have started to consume imported manufactured goods (transistor radios, refrigerators, furniture, dark suits, preserved food, Scotch whisky etc.). The prices of imported goods have risen for a number of years. This means that the landlords—especially the small and medium ones—in order to maintain their status and standard of living, must increase the exactions from their tenants. And the exactions must be convertible into cash.[39]

In this way the tenants are indirectly affected by the international capitalist market. Their cash-crops (wheat, teff, noug, pepper) are not exported, they are brought to Addis Ababa and consumed there. But the landlords increase the exactions from the tenants to be able to furnish themselves with imported goods.

Those who own larger areas can choose to introduce mechanized farming. The Government policy for promoting mechanized production has mobilized an increasing number of landlords and financially strong urban residents for the 'green revolution'. These people have realized that money can grow fast on mechanized, fertilized, subsidized fields. In Bako the commercial farmers are not from the local landlord class, but if their enterprises turn out to be profitable, it is reasonable to suppose that the landlords in Bako, or at least their educated children, will take a more 'innovative' attitude to agricultural production.

The spread of mechanization is also dependent on the network of roads and the local geography. Communication between Addis Ababa and Bako is good, the main road allows trucks to make the journey in one day. Communication to the region north of Bako is also improving. In 1971 SIDA in co-operation with the local population started a labour intensive road building project from Bako to Jere, 11 km to the north. This road was opened for motor traffic in 1973 and will be continued 50 km north to Shambu, capital of Horo & Gudru *awradja*[40]. This *awradja* comprises a fertile highland of softly rolling hills. The social structure of agricultural society there is similar to Bako; 67 % of the holdings in Horo & Gudru are operated by tenants.[41]

The road project has been welcomed by all, this is expressed by the fact that 70 % of the costs for the road project has been collected from the local population and the remainder has been contributed by SIDA. Swedish peace corps volunteers supervised the road construction. The road may of course bring many advantages to the population. This should nevertheless not obscure the fact that, given the present social and economic structure of agricultural society and given the prevailing trend towards mechanization, the road might open up the area for the same kind of agricultural development seen in Sheboka: the entry of tractors and harvesters into the fertile lands and the exit of redundant tenants.

The conclusion of the above argumentation is that Bako minimum package project has not been able to challenge the feudal structure and assist its own target group. Instead, a commencing capitalism in the form of mechanized farming is beginning to break up feudalism. In this process the package project stands powerless in the face of an acute crisis that the peasant society is undergoing.

Notes to Chapter 8

[1] *Survey of Shoa,* p. 25; *Survey of Wollega,* p. 25.

[2] *Survey of Shoa,* p. 27; *Survey of Wollega,* p. 27.

[3] *Survey of Shoa, p. 23; Survey of Wollega,* p. 22.

[4] *Survey of Shoa,* p. 27; *Survey of Wollega,* p. 26.

[5] *Report on Land Tenure Survey of Welega Province,* p. 15.

[6] *Ibid.,* p. 13.

[7] *Survey of Shoa,* p. 30; *Survey of Wollega,* pp. 30 f.

[8] *Land Tenure Survey of Welega Province,* p. 44.

[9] E. Kelber–Awole Mela, *Agricultural Survey of Bako Area* (1970).

[10] Taye Gulilat–Guenter Abraham, *Socio Economic Study of the Bako-Jere Area* (1973).

[11] *Ibid.,* p. 19.

[12] *Ibid.,* p. 23.

[13] Berhanu Gebre Mariam, *Adoption of Innovation at Bako Minimum Package Project* (1973), p. 12. According to Berhanu an estimated 4,000 tenants and 300 landowners were recorded by the *woreda* governor within Gebu Seye *mikitil woreda* (where the extension areas Kejjo and Anno are located) and on 120 *gashas* in Sheboka there were 554 peasant households out of which 482 were tenants.

[14] Information obtained during 1971–73 from a number of interviews and discussions with persons well acquainted with Bako area.

[15] Berhanu Gebre Mariam (1973), *op.cit.,* p. 13.

[16] *Ibid.,* p. 14.

[17] *Ibid.*

[18] *Ibid.*

[19] Mulugeta Tessemma, *Land Use Study in Lege Batu* (1971).

[20] Shiferaw Melesew, *A Land Use Study of Botche* (1971).

[21] Tefera Bitima, *Land Use Study of Facha* (1971).

[22] Information obtained from the project office in Bako.

[23] Taye Gulilat–Abraham, *op. cit.,* Annex I.

[24] The figures differ on some points from the figures presented in EPID's annual reports. The most conspicuous difference is the total number of participants (cash + credit) in 1973. EPID's annual report gives a total of 304, while the figure I obtained from the project office is 223. This difference can partly be explained by the fact that EPID's calculation is based on the number of sales reported (one farmer can make more than one purchase) while the figure I received from the project office is based on the participating individuals.

[25] Information obtained from the project office in Bako and from the Evangelical Church Mekane Yesus in Bako. The settlers' credits are included in the figure for Anno.

[26] Information obtained from the project office in Bako.

[27] *Ibid.*

[28] *Ibid.*

[29] *Statistical Abstract 1971*, p. 120.

[30] Information obtained from the project office in Bako.

[31] *Ibid.*

[32] *Ibid.*

[33] *Ibid.*

[34] Cf. note 14.

[35] Cf. note 14.

[36] Taye Gulilat-Abraham, *op. cit.*, p. 24.

[37] Cf. note 14.

[38] Asfaw Edessa, *A Micro-Geographical Study of Mankata* (1970), pp. 61 f.

[39] For this point I am indebted to discussions with Siegfried Pausewang and Jan Olof Lundberg.

[40] Thomas Carlsson—Arne Carlsgårdh, *Road Construction Project Bako-Jere Final Report* (1973); *Bistånd Genom SIDA 1972/73 Verksamhetsberättelse*, p. 68.

[41] *Survey of Wollega*, p. 29.

Chapter 9 Shashamane Minimum Package Project

9.1. An Outline of the Social Structure

General Information

Shashamane minimum package project started its activities in 1971. The project centre is situated in the town of Shashamane, 200 km south of Addis Ababa on the main road to Kenya. Administratively the area is part of Haikoch & Butajira, the southernmost *awradja* in Shoa province. Geographically the area lies in the Lakes' Region in Rift Valley at a general elevation of 1,500–2,000 meters. The profile of the landscape is a softly rolling acacia savannah. The population is predominantly of Oromo stock, with elements of Sidamas and Amharas interspersed. Both the Ethiopian Orthodox Church and Islam have numerous followers.

History

During the nineteenth century the Lakes Region was a border land between pastoral Oromo tribes and small chieftaincies and kingdoms composed of agricultural Oromo, Sidama and Gurage groupings. After a series of military campaigns in the 1880's and 1890's the whole region came under Menilek's rule. Structural change after the conquest followed the same pattern as elsewhere, the land being appropriated by the Crown and the usufruct of it given to military commanders and subjugated co-operative rulers. The pastoralists were pushed out and agriculture gained terrain. The new landlords invited peasants from other areas to settle down on their holdings as tenants. Large areas remained grazing land.

Land Tenure Pattern in the 1960's

In 1971 and 1972 the project's activities were concentrated to the two extension areas Shashamane and Negele*, within Haikoch & Butajira *awradja*. Some data from CSO's sample survey carried out in the early 1960's, will be presented as an introduction.

Around 13 % of the total area of the *awradja* was under cultivation (some 144,000 hectares)[1]. Cereals dominated agricultural production,

* Although there were two additional extension areas in Leku and Yirgalem in Sidamo province.

maize being the most important crop.[2] Agriculture was practised in small family operated units, 89 % of the peasant holdings comprised 3 hectares or less. Tenancy was common: 47 % of the peasant holdings being operated by tenants, 7 % by farmers who owned some fields and rented others and 46 % by owner-cultivators.[3]

The total number of holdings was estimated to be 87,000[4]. In the project area the percentage of tenant holdings was even greater. The *woreda* governor's office in Shashamane had 2,879 owners and 15,843 tenants registered in Shashamane in 1970.[5] Many, although how many is not precisely known, of the owners were absentee landlords. For all of Shoa province, it was estimated that 45 % of the measured area was under absentee ownership.[6]

A slight majority of the tenants paid rent in the form of a fixed sum of money, while 45 % were share-croppers. Among the latter the rent varied between 25 % and 50 % of the harvest, provided that the tenants used farm implements and work oxen of their own.[7]

This feudal pattern is now in a process of change, a process which began in the mid 1960's and accelerated at the turn of the decade. The trend is towards large scale commercial farming. Shashamane and Negele *woredas* have become centres in this process, which is a result of the Goverment's policy of encouraging large scale mechanized farms as a vehicle of development. Since the package project has extension/marketing areas in both *woredas*, a recapitulation of the events there leading to the mechanization of agricultural production will be sketched below.

Starting in 1966 FAO in co-operation with the Ethiopian authorities chose a number of places in the countryside for agricultural experiments with the purpose of identifying the effects of fertilizer in the areas. Shashamane and Negele were among the selected places. The project was called "Freedom From Hunger Campaign" (FFHC) and included a credit programme for fertilizer.[8] In 1968 FFCH started credit operations in Shashamane.

In practice the campaign addressed itself to large landowners, i.e. persons who could provide security for credits and who had large enough land to make tractor operated fertilized production profitable.

The FAO project gave fertilizer credits and the AID Bank gave credits for the purchase of tractors and other agricultural machinery.

The existence of a domestic market demand for cereals in the towns made a number of wealthy people respond to the Government sponsored credits and subventions and drew them into agricultural production. The cash crops grown have been wheat, maize, teff and, since 1973, haricot beans.[9]

The rapid increase in the establishment of commercial farms can be inferred from the following table.

TABLE 9. Number of agricultural machines in Negele *woreda.*

Year	Tractors	Harvesters
1968	3	1
1969	5	3
1970	19	5
1971	37	12

Source: The project office in Negele.

The progressive increase has continued up to 1974. In late 1973 the commercial, tractor operated farms in Shashamane and Negele were estimated to number between 100 and 150. The largest farm comprised about 30 cultivated *gashas* while the median size was 2 to 5 *gashas.*[10] Due to the commercial farms, Shashamane has grown to a bustling town, inhabited by about 14,000 persons in 1973.[11] In addition to the abundant occurrence of bars in Shashamane there are numerous gas-stations, garages, tyre repair shops, sparepart shops for agricultural machinery etc. and more are under construction (1973). Furthermore, Shashamane is strategically located as a traffic junction for southern Ethiopia. The all weather road to Kenya which is to be completed in 1975 stretches through the town. Units of the Imperial army are stationed close by.

It was noted that the newly established commercial farmers were wealthy individuals who could provide security in cash or in land for the FFHC credits. Actually, the farmers can be divided into two subgroups: i) individuals who owned land in the Shashamane-Negele area prior to the credit programme. These are either descendants of Menilek's grantees or have received grants from Haile Selassie, ii) individuals who did not own land in the area prior to the mechanization process, but who bought or contracted land when it became known that large scale commercial farming would pay off in Shashamane-Negele. In the case of contractors the rule is that they contract an area from a large landlord or from the Government for a period of five to fifteen years. They pay cash for the contract and can thereafter use the land according to their plans. The two sub-groups have an important feature in common, they belong to the already privileged stratum around the Imperial Government, many of them are retired army officers or civil servants.

One landowner in the area is superior to the others. This is the Imperial family.[12] The heirs of the late prince Sahle Selassie own almost one half of the area of Negele *woreda*. The land is administered by a local representative on behalf of the heirs. The majority of the tenants in Negele are Imperial tenants. Until 1970 the Imperial land was used in the traditional way; parts of it lay unutilized, parts were grazing land and parts were cultivated by tenants. That year, however, mechanized commercial production was introduced on the Imperial land. In 1971 about three hundred hectares were put under mechanized cultivation and the farm is expected to expand by about one hundred hectares yearly. For this reason a farm-centre has been established in Negele, with an office, storage facilities and machinery.

As elsewhere in the highlands mechanization coincides with eviction. No detailed study has, to my knowledge, been made of the frequency of evictions in Shashamane-Negele, but initiated persons estimate the rate of eviction to have been between 100 and 200 families yearly since 1968. In the agricultural area Negele-Shashamane-Awassa the number of evicted tenant families was estimated at approximately 200 in the years 1971 and 1972.[13]* The mechanization does not only affect already cultivated land, it also affects traditional grazing land. This means that some of the evicted tenant families have lost the possibility to cultivate as well as the possibility to graze their cattle. These families have been forced to leave the region. It is reported that they have moved to the nearby lakes with their livestock. The influx of people to the shores has given the landowners there new sources of income—they charge the evicted tenants rent for grazing their cattle there.

The practise of eviction in the process of mechanization is approximately this: when the landlord has made up his mind for commercial agriculture, the tenancy relationships in the area to be affected by mechanization are terminated. Later on the tractors arrive. They plough up the land. The home of the tenant family is generally situated in the immediate proximity of the fields. The tractors plough up the soil around the house and the tenant family is told that it is prohibited for people and cattle to walk in the furrows which extend right up to the house. The family has no choice but to abandon their home.

The breaking up of tenant families is a consequence of eviction. Parents and infants move around in the countryside with their livestock in the hope of finding land to rent. Teenagers prefer to make their fortunes in the towns, where they form part of the growing number of coolies, street vendors and prostitutes. The process of eviction has of course many variations, and extreme brutality has been shown in some cases. Here it is

* This is equivalent to between 500 and 1,000 individuals a year or between 3,500 and 7,000 individuals for the period 1968—1974.

sufficient to add that many tenants have built their homes in hilly and stony places, which under no circumstances are suitable for mechanized agriculture. Even though the fields of these tenants have been taken over by tractors their houses are still intact and maybe even some plots of land where they can cultivate subsistence crops. Under such conditions the tenants have not been forced to leave the area. Instead they form a labour reserve which is used by the commercial farmers during weeding and harvesting time. Thus, their status has changed from peasants to seasonally employed agricultural labourers. A feudal system of peasant exploitation has thus been transformed to a capitalist system of exploitation.

Some tenants have, at least temporarily, managed to adapt to the process of mechanization. In the immediate neighbourhood of Negele most of the holdings, including tenant holdings, are machine operated.[14] A number of tenants in a neighbouring area join together and rent agricultural machine-service from their landlord, i.e. the ploughing and harvesting of the tenants' fields are performed by the landlord/commercial farmer's hired workers and machine pool. In exchange, the tenants pledge to sell their harvest to the landlord.

Actually, the transactions are made in amounts of grain, the value of which is converted into money by the landlord who calculates with the prevailing market prices. First the landlord appropriates an amount of the tenants' harvest that is equivalent to the rent, then he takes an additional part which is compensation for the machine-service. For the remaining part of the harvest the landlord pays his tenants. Since the market prices which are used in the calculations are at their lowest during harvest time, it is the landlord who benefits from these transactions. The market mechanisms are used to exploit the tenants.

Further information on how mechanization changes the conditions of the peasantry is provided in a study of a village close to the minimum package project in Awassa *woreda*. The area studied comprised 22 hectares on the southern shore of lake Awassa.[15] This area was at the time of the study (1971) cultivated by ten tenant families. The leaser of the land was a commercial farmer who altogether had 200 hectares under mechanized cultivation. He had contracted the land from its owner, the Imperial Government. The 22 hectares studied were ploughed and threshed with the machinery of the commercial farmer by his hired, skilled workers. The commercial farmer decided alone what crops were to be grown. The role of the ten tenant families was confined to performing the tasks of sowing, weeding, harvesting and transporting the crop (maize) to the store of the commercial farmer. The tenant families were responsible for an area of one to four hectares each. Four of the tenants worked furthermore as part time labourers for the commercial farmer, assisting his tractor drivers. The tenants had written lease agreements with their landlord. According to these agreements 60 % of the harvest went to the landlord/commercial farmer while 40 % was retained by the tenant.

The study referred to above seems to describe a transitional stage between feudalism and capitalism. It is not difficult to foresee that the commercial farmer can invest in more machinery which makes it even more profitable for him to hire workers who for a fixed cash salary perform all agricultural work. Some of the present tenants may then be employed (those who already are part-time workers) while the rest become superfluous and must leave.

The landlords who have gone over to commercial farming prefer capital intensive methods rather than labour intensive methods in agricultural production. This is reflected in the following statement made by a commercial farmer in Wondo-Gennet, close to Shashamane:

> "I have 1,300 tenants. Now I am cultivating the land by myself. I employed the tenants for manual labour, weeding, farming, harvesting but found them incompetent. Therefore, I have decided to use improved mechanization . . ."[16]

This preference for mechanization is a direct result of the Government's agricultural policy. A few comments on the long-term consequences of continued expansion of large scale mechanized farming will be summarized below.[17]

The commercial farmers practise extensive farming. Their interest is not directed to maximising the yield per hectare, but the total yield. In Shashamane-Negele the combination of large tracts of land (hitherto cultivated by tenants or used for grazing) which are suitable for mechanization and the relatively inexpensive inputs (subsidized agricultural machinery, fuel and fertilizer and underpaid daily workers) invite commercial, mechanized farming. The fertilizer increases the yields and the use of tractors makes it possible to cultivate large areas. The Imperial Government has prepared this situation through the land grants and through the subsidies to mechanization.

It has been pointed out that the major reason for the profitability of mechanization does not lie in the per hectare profit but in the marked differences in accessability to information, credit and factors of production between the poor peasants and the commercial farmers.[18] Without Government subsidies* and without the power to evict tenants and employ

* The duty-free tractor fuel was abolished in 1973, but agricultural machinery could still be imported duty-free.

some of them as daily workers for minimal salaries, large scale mechanized farming would not pay off for the individual farmers. For the Ethiopian national economy, mechanized farming does not pay, since the main inputs must be imported. And for the poor peasants mechanization is directly disastrous.

It is known that fertilizer increases the yield only in the short run. If no organic matter is restored to the soil, it will be exhausted after some ten years, however much fertilizer is spread on it. The present commercial farmers in Shashamane-Negele are not agronomists, nor do they consider long term agronomic effects on the soil. Instead their attitude towards farming is purely business oriented, they think in terms of maximum profitability within the shortest possible time. (Many of the farmers reside in Addis Ababa, visiting their farms only on weekends). Therefore the present practise of the commercial farmers is to plough up extensive areas and scoop out fertilizer. In the long run this robber exploitation will be disastrous for the country. It maximises eviction and destroys the soil. If the present practise is allowed to continue, it is not unlikely that the Shashamane-Negele area in the end of the 1970's will be exhausted and the commercial farmers will abandon it for virgin lands further south. This area will by then have been opened up for motor traffic by the World Bank and USAID supported highway programme. And the process eviction-mechanization-depletion of soil will be repeated.

One presumed positive effect of large scale mechanization is the fast increase in food stuffs available for domestic consumption. The increase in production has reduced the import of wheat and wheat flour. But this increase in production has not benefitted the population. Surplus production has been bought up by large merchants and stored for speculative reasons. The result has been increases in consumer prices.[19]

The future expansion of this kind of commercial agriculture is dependent on monetary demand for crops which the commercial farmers can cultivate. Up to 1973 the production was geared to consumption in Ethiopian towns. It functioned as import substitution since Ethiopia had earlier bought much wheat and wheat flour from abroad. But the monetary demand in the towns is limited and temporary oversupply caused prices to fall in 1972. In 1973 cereal prices were again moving upwards. These price fluctuations influence commercial farmers' investment decisions and thereby the rate of expansion of mechanized agriculture in the short run. If the commercial farmers manage to get a foothold for their crops on the international market (which seems to be the case for haricot beans) an increased tempo in the process of mechanization and eviction is to be foreseen.

So far the discussion has concerned the monetary demand for crops. In reality there is an enormous need for cereals within Ethiopia. In the summer of 1974 there were famine conditions in parts of Tigre, Wollo, Shoa, Gemu Gofa, Bale and Hararge provinces. But the starving masses, amounting to several million people, cannot express their need in monetary terms. Consequently, the commercial farmers do not consider their demand unless international relief organizations interfere and buy from them.

9.2. The Project's Field Activities

Coverage

In 1971 Shashamane minimum package project had extension areas/marketing centres in Shashamane, Negele, Leku and Yirgalem. Only the activities in Shashamane and Negele will be considered here.*

In late 1971 the project area comprised an area around Shashamane town with a radius of 18 km and in Negele an area 10 km x 20 km along the main road to Addis Ababa. One extension agent and one marketing assistant were stationed at each of the extension areas/marketing centres.

In addition to the project's demonstration fields there were seven model farmers in Shashamane and four in Negele. These were owner–cultivators chosen by the neighbouring peasants.[20]

The Credit Programme

The demonstration of the effects of fertilizer and the provision of inputs on credit were the most important activities of the project in 1971 and 1972. The following table shows the distribution of credits and cash sales.

TABLE 10. Shashamane minimum package project fertilizer distribution 1971 and 1972. No. of participants.

Extension area/ marketing centre	1971			1972		
		credit			credit	
	cash	owner/ cultiva-tors	tenants	cash	owner/ cultiva-tors	tenants
Shashamane	47	186	117	184	117	130
Negele	55	163	215	65	232	242
Leku	–	12	1	8	109	4
Yirgalem	14	119	–	62	208	–
Total	116	480	333	319	666	376

Source: The project offices in Shashamane and Negele.[21]

* Yirgalem has a special character since it comprises a settlement. In Leku the project activities have been "insignificant" according to the project staff.

The total number of participants should be compared with the estimated 10,000 peasant households in the project area. The total number of participants is of a different size than in Bako. This must partly be explained by the general commercialization of agriculture in Shashamane-Negele. Because of the FAO project the effects of fertilizer were already known in the area when the minimum package project entered the scene. Considering this, the actual number of participants could have been still higher. Again, it is credit regulations that restrict participation among the poorest stratum of peasants. This is clearly indicated in a report from the project supervisor:

> "The inability of paying the downpayment is another problem which has prevented an unknown number of convinced peasants from participating in the programme".[22]

An analysis of the different categories of participants shows that the cash buyers have increased most rapidly, in 1972 they comprised 1/4 of the participants. When it comes to the credit takers here classified as owner-cultivators it must be mentioned that 23 of them in 1971 and at least 59 in 1972 were landlords, cultivating up to 20 hectares (the upper limit for participation) and leasing out additional land to tenants. The number of tenants seems amazingly high compared to Bako. In Shashamane and Negele tenant participants approximate 50 % of the credit takers. It has, however, been reported that in a number of cases the tenants act only as instruments for their landlords, on whose fields the fertilizer is used.[23] Nevertheless, there are a number of tenants who have used fertilizer on their own fields. But it is difficult to know to what extent these tenants have themselves benefitted from the credits. A mathematical cost-benefit analysis which EPID uses is not illuminating (cf. chapter 10). Landlords might approve of their tenants' use of fertilizer with the explicit aim of increasing rents.

Actually, the credit takers' benefits in cash terms from the credits 1971 and 1972 were minimal. As mentioned earlier, the prices of cereals and especially maize, the main crop in Shashamane-Negele, were extremely low in 1972. The peasants were obliged to sell shortly after harvest because of their limited storing capacity and because taxes as well as rent are payable at that time. The low prices made repayment of credits difficult (EPID does not accept repayment in kind). As to the credits given in 1971, 120 farmers had not repaid their debt at the due date in 1972, but many paid later in the autumn. Still, there were 46 defaulters left in 1973 and EPID took legal action against them.[24] The number of defaulters was even greater concerning the credits given in 1972. One year later 95 credit takers in Shashamane and 173 in Negele had failed to pay.[25] As this number was far too high for EPID to tolerate, it was decided to discontinue the credit programme in the Shashamane-Negele area. No credits were given there in 1973.[26] Legal action was taken against the defaulters, most of whom were tenants. Those who cannot pay when they are brought to court will go to prison.

The credit programme of the Shashamane minimum package project did not live up to the expectations and officially stated goals. On the contrary, a considerable number of medium landlords and rich peasants were assisted by the programme. Poor peasants were also assisted but as price fluctuations hit them severely, many ended up in a worse economic situation than before they participated in the credit programme and some will even end up in prison.

Co-operative Promotion

In 1973 there were two registered co-operatives in Shashamane and one, not yet registered, in Negele. The minimum package project has been involved in the formation of these, but has not been the prime motivator. The co-operatives are dominated by commercial farmers, who have realized that co-operatives can constitute a proper organizational framework for the promotion of their business interests. This will be revealed by a closer look at the co-operatives.

Shashamane: The co-operative society in Melka Odekuyera was registered in 1971, and had some 200 members.[27] The activities of the co-operative are purely business oriented, the co-operative obtaining fertilizer credits and cash loans from the AID Bank. The fertilizer is distributed to the members and the co-operative buys up the harvest of the members. The grains purchased in this manner are sold on the national market. The executive committee of Melka Odekuyera was in 1973 dominated by commercial farmers, the president being a retired colonel. Among the members there were many small owner-cultivators and tenants.

The other co-operative, Aja, had in 1973 a little more than 200 members.[28] The activities were similar: the society functioned as a credit link between the AID Bank and the members, and as a marketing link between members and the market. The executive committee was dominated by commercial farmers/merchants, the president being a high official within the Ethiopian Grain Board.

Negele: In 1971 the minimum package project assisted in forming a preco-operative.[29] The executive committee was infiltrated by commercial farmers/merchants. Somewhat more than 100 members paid registration fees and shares. The president and a few other members of the executive committee were merchants of such magnitude that together they controlled the whole grain market in Negele. This concentration of power within the society became known and obviously made small farmers reluctant to join the co-operative. The merchants in the executive committee tried to use their position to promote their private business interests. Since they could influence the grain prices in Negele at harvest time, they tried to buy the members' (and other farmers') harvest at a cheap price, sell it for better prices later on in Addis Ababa and render deficient accounts to the members. The minimum package project staff tried to put pressure on the president to withdraw, and refused to hand in papers for proper registration of the co-operative to the Ministry of National Community Development unless the executive committee was reorganized. In 1973 the president and a few others resigned from their posts since the members were passive and the co-operative showed little activity. The reorganized executive committee took the following form: the president, vice-president and secretary were commercial farmers cultivating 1—2 *gashas*, the assistant secretary was foreman for the workers loading grain on the trucks at Negele market, and the treasurer was a retired court official, who had taken up commerical farming and cultivated 1 1/2 *gashas*.

One impression of this brief overview is that the agrarian co-operatives in Shashamane-Negele are not organizations for small peasants but for commercial farmers. The real function of these organizations is to serve as marketing channels for large farms. The co-operatives do have a number of members that are small owner-cultivators and tenants, but the co-operatives cannot be said to be *their* organizations. The inferiority of the poor peasants, especially the tenants, in their social relations to the landlords/commercial farmers/merchants and their ignorance of accounting, reading documents and so forth make them vulnerable for manipulation. It is also openly admitted by EPID's officials that commercial farmers join the co-operatives in order to promote their private interests, which do not coincide with the interests of the poor peasants. The social inferiority of the poor peasants makes it easy for the commercial farmers to monopolize the executive committees.* The legal framework of co-operatives (cf. chapter 10) opens the door for landlords, merchants and large scale commercial farmers to membership in co-operative societies.

At the time of my visit in 1973 the project staff did not consider to encourage special co-operative societies exclusively for poor peasants. Instead, their policy consisted of trying to persuade the poor peasants to

* Positions in the executive committee are subject to election by the members and the poor peasants vote for people with status and wealth.

join the already established co-operatives. I was told that this policy showed little success since many of the peasants seemed to ignore this message from the project staff.

9.3. SORADEP

A few words must be said about the existence of institutions parallell to the minimum package project. In Awassa, 20 km south of Shashamane, the Ministry of National Community Development has established a project with similar aims as EPID–to raise the economic standard of the rural population and to make social development possible.[30] This project is called The Southern Region's Agricultural Development Project (SORADEP). Its activities cover parts of Awassa *awradja* as well as Shashamane and Negele *woredas* in Haikoch & Butajira *awradja*. An approximative 60,000 peasant families, mostly tenants, cultivating 0.8 to 5 hectares, live in the project area, and there are numerous large scale mechanized farms.[31] SORADEP gives credit for seeds (maize) and fertilizer, and thus overlaps the activities of Shashamane minimum package project.

The activities of SORADEP are illustrated in the following table.

TABLE 11. Credits extended by SORADEP

No. of farmers taking credits for:	1971	1972	1973
Maize-seeds	1,650	6,021	13,127
Fertilizer	–	942	2,127

Source: Borderon, *Introduction to SORADEP*, p. 14.

SORADEP has been more successful in reaching the peasantry than has the minimum package project, and many of SORADEP's participants are found in the area where the two projects overlap. The reason for SORADEP's results can be found in its credit regulations. Credits are given only to small farmers, but no lease agreement with the landlord is required for tenants, and the downpayment is 8 % of the credit value.[32] The model-farmer system is not used. Instead, SORADEP gives short courses in agriculture for young villagers who have finished school and employ them in their home villages as supervisors. In 1973 the project had 93 agricultural advisors in its project area. Thus the project strives to exercise "agronomic security" over the credits.[33]

As the table above shows this approach to the credit programme has been able to reach far more peasants than EPID's approach (formal securities, high downpayment and instruction through a few model-farmers).

Thus, SORADEP now appears to take over credit activities among the peasants on the ruins of Shashamane minimum package project.

9.4. Conclusions

In spite of its different approach, SORADEP faces the same problems as the minimum package project when it comes to realization of the goal 'social development'. Basically both projects think that an increased production must come first, thereby enhancing the economic situation of their target group. Later on activities towards social improvements can be undertaken. This policy might be reasonable as such, but within the class society of rural southern Ethiopia where economic and political power is concentrated in the hands of landlords, merchants, commercial farmers, and local governors, the policy does not work.

The result of SORADEP's activities so far has been an increase in the *production* of the credit takers.[34] But this is not the same as an improved standard of living. The majority of the tenants pay rent in cash and they are required to pay at harvest time. Their increased production makes however the prices go down, which means that a considerable part of the increase goes to the landlords. Many of the landlords are themselves merchants and commercial farmers, in such cases the tenants actually pay in kind, i.e. the landlord calculates how many quintals of grain are equivalent to the agreed cash rent. Since the landlords use the very low price which is current at harvest time in their calculation, the amount of grain paid is considerable. Furthermore, it is well known that the formal cash rents are continually increased when the tenants start to use fertilizer. SORADEP and the minimum package project assist poor peasants to increase their production which signifies an increased market supply and, consequently, lower prices at harvest time. Thus it can be argued that under prevailing conditions the provision of fertilizer to poor peasants only serves to intensify the exploitation of them. Neither SORADEP nor the minimum package project provide any *political* support to their target group. They stand powerless when confronted with the process of large scale commercial farming and the landlords continuing and modernized exploitation of their tenants.

Some comparisons betwen Bako and Shashamane can now be made.

In Bako, landlords want to restrict their tenants' use of fertilizer, while in Shashamane landlords encourage tenants to use fertilizer. In Bako landlords view the formation of co-operatives with suspicion; in Shashamane landlords are eager to join co-operatives.

The reasons for this difference in attitude must be looked for in the changing organization of agricultural production. The rural society around Bako is still predominantly characterized by a feudal mode or production. The landlord class enriches itself by collecting tribute from the individual tenant families. In the Shashamane area, agriculture has been rapidly penetrated by the capitalist mode of production. The former feudal landlords have changed into or been replaced by commercial entrepreneurs. The establishment of large farms has brutally destroyed the feudal relations between landlords and tenants. The producers are being evicted, i.e. separated from their means of production. The members of the evicted families reappear in the guise of daily workers, prostitutes, beggars and thieves. Since only a limited number of them can obtain regular employment, the process is, in Amin's terminology, one of 'marginalization'.[35]

Although the capitalist mode of production is rapidly expanding in the Shashamane-Negele-Awassa area, it is not yet dominant in Haikoch & Butajira *awradja*. Feudalism continues to exist, but the tribute is becoming closely related to the commercialization of agricultural production. If the profitability of cash crops like maize, wheat, beans and oil seeds continues, the capitalist mode of production will continue to expand and a growing number of poor peasants will be evicted. Shashamane shows Bako its own future.

A development similar to the one in Shashamane has been analysed by Hamza Alavi in Punjab in Pakistan. In the areas studied by Alavi landlords have turned to mechanized farming and caused evictions among the tenants who were earlier share-croppers.[36] Alavi emphasizes that this process does not mean that the capitalist mode of production replaces the feudal mode. Instead, both modes coexist since

> "the owners have also retained sharecroppers, taking away only a part of their land for mechanized cultivation. They keep the sharecroppers because they need them, as a source of seasonal labour."[37]

Alavi concludes that one might speak of a 'multiplex' mode of production in this case.

The situations in Punjab and southern Ethiopia are structurally similar. An indispensable prerequisite for the profitability of large scale commercial farming is the existence of cheap labour. This cheap labour force is provided by the subordinated peasants who, due to the encroachement of mechanization, no longer can or are allowed to earn their subsistence through traditional farming.

I find it fruitful to use Amins terminology on the 'articulation' of modes of production in order to shed theoretical light on this situation. The capitalist mode of production is articulated, i.e. linked, to the feudal mode of production. In this case the result, in terms of the consequences for the peasants, has been a change in the mechanisms of exploitation.

Finally, what is the role of the minimum package project in this process? First, the experiences from the co-operative promotion programme show that the project does not try to provide an organizational framwork for the interests of the poor peasants. Second, the project, like SORADEP, contributes to the encouragement of commercialization of agricultural production through its emphasis on the use of fertilizer. Since the peasant producers have no organization of their own this means only an increased exploitation of them by landlords and merchants.

Third, the project may have assisted a number of middle and rich peasants, such who are not eligible for credits from the AID Bank but who still are in the upper stratum of the peasantry, just under the landlords. By helping them the project is creating a group of kulaks.

Fourth, it is plausible to believe that the project definitely destroyed its possibilities for winning the confidence of the poor peasants when it started threatening to bring a number of them to court.

Notes to chapter 9

[1] *Survey of Shoa,* p. 25.

[2] *Ibid.,* p. 23.

[3] *Ibid.,* pp. 27, 29.

[4] *Ibid.*

[5] Berhanu Gebre Mariam, Shashamane and the Agricultural Development Project. in *HSIU and USAID Summer Interns* (1970), p. 55, footnote 7.

[6] Varghese, *op.cit.,* p. 74.

[7] *Survey of Shoa,* p. 30.

[8] FAO, *Freedom From Hunger Campaign, Fertilizer Programme, Near East and North Africa* (1968).

[9] Information obtained from the project office in Shashamane.

[10] Ibid.

[11] The population figure for Shashamane given by CSO, *Urbanization in Ethiopia, Statistical Bulletin No. 9* (1972) is 10,890 in 1970. Assuming a growth rate of 9 % per annum the figure for 1973 is around 14,000.

[12] Information obtained during 1971–73 from a number of interviews and discussions with persons well acquainted with the Shashamane-Negele area.
[13] Cf. note 12.
[14] Information obtained from the project office in Negele.
[15] Kebede Tessemma, *Land Use Study in Loke* (1972).
[16] Quoted from Berhanu Gebre Mariam (1970), op. cit., p. 57.
[17] Cf. note 12.
[18] Gene Ellis, *Man or Machine Beast or Burden, A Case Study of the Economics of Agricultural Mechanization in Ada District, Ethiopia* (1972).
[19] I quote the following discussion of manipulation of market prices in Ethiopia from an unpublished manuscript by Lars Bondestam:

> "Despite a production increase, the retail price index for food for Addis Ababa reached in May 1971 a level higher than ever before (*Retail Price Index for Addis Ababa,* a monthly information published by IEG CSO). There has been at least two attempts to explain this increase in food prices on the Addis Ababa market. Both explanations suggest that the increase was due to an unbalanced supply-demand, which is contradicted by the fact that production is increasing and that the total purchasing power of the Addis Ababa-citizens is probably not increasing faster than its population grows (some 6–7 % yearly). The conclusion in the one of the papers (Berhanu Disasa, *Ethiopa's Agricultural Situation—a Hypothesis,* CSO, June 9, 1971) is based mainly on interviews with traders, which is a quite unscientific approach bearing in mind that these people are the links between production and consumption. The other, more serious paper (T. James Goering, *A Note on Food Prices*, Planning Commission Office, Agriculture Division, June 24, 1971), suggests that "the major causes for cereal price increases are pervasive supply-demand unbalance". Now afterwards, we now that the reasons were quite different: the increase in retail prices in Addis Ababa were mainly due to artificially prices set by the traders. The paper also touches this point when stating that "the role of the Ethiopian Grain Corporation in the determination of cereal prices is not clear", but "it operates as does any large trader, with the making of an *annual* profit an important consideration. This objective could at times conflict with the price stabilization objective". The succeeding event in June 1971, when secondary school students forced a sudden decrease in the prices of some products (by mixing sand in the butter and threatening to burn up the grain stores of the Grain Corporation), also justifies the belief that the prices had been artificial rather than a "natural effect" of the supply-demand situation. Probably for the first time in Addis Ababa's history, the retail price index for cereals went down from May to September by as much as 10 % (and yet, this was before the harvest—during this time prices usually go up!)."

[20] Information obtained from the project offices in Shashamane and Negele.
[21] The figures differ on some points from the figures presented in EPID, *Annual Report for 1971/72 Financial Year.* The annual report has 606 owner-cultivators and 207 tenants taking credit in 1971 and 687 owner-cultivators and 381 tenants taking credit in 1972.
[22] Report from the project supervisor to EPID's headquarters, dated July 30th 1971.
[23] Information obtained from the project office in Shashamane and from EPID's headquarters. For details, see Michael Ståhl, *Contradictions in Agricultural Development, A Study of Three Minimum Package Projects in Southern Ethiopia* (1973), pp. 41 f.
[24] Information obtained from the project offices in Shashamane and Negele.
[25] *Ibid.*

[26] *Ibid.*
[27] *Ibid.*
[28] *Ibid.*
[29] *Ibid.*
[30] Alain Borderon, *Introduction to the Southern Region's Agricultural Development Project* (1973), p. 6.
[31] *Ibid.*, p. 3.
[32] *Ibid.*, p. 11.
[33] *Ibid.*, pp. 6 ff.
[34] *Ibid.*, p. 15.
[35] For a discussion of the concept of 'marginalization' see Amin (1973), *op. cit.*
[36] Hamza Alavi, Peasant Classes and Primordial Loyalties, in *The Journal of Peasant Studies,* Vol. 1, No. 1, 1973, p. 53.
[37] *Ibid.*

Chapter 10 Further Discussion of the Minimum Package Projects

In this chapter the scope of the analysis will be broadened to cover the credit programme and the co-operative promotion programme in general in the minimum package projects which operate in the provinces where private ownership of land prevails. The analysis will lead up to some tentative conclusions on the reformistic capacity of EPID regarding its main goal of participation.

It should be pointed out that this analysis does not claim to be an evaluation of EPID as a whole, but focuses on the structural possibilities of EPID to assist the peasantry, and especially the tenants, in the southern provinces to assume responsibility for the activities aiming at an improved standard of living.

10.1. The Credit Programme

During the period 1971—73 nine to ten minimum package projects were launched annually, and for 1974 twenty new projects were under preparation. In 1974 the number of projects will be 48 and, in theory, cover half a million peasant households.[1] Up to 1973 the most conspicuous activity of the projects was their credit programme. Credits have been given for seeds and fertilizer. Relatively few farmers have, however, taken credits for seeds; the overwhelming majority of credits has gone to fertilizer.[2] Table 12 illustrates numerically the distribution of credits to farmers. Cash buyers have also been included. The projects within the communally owned areas and the privately owned areas are grouped separately.

Considering the projects in the privately owned areas, a general feature is that the projects with the highest rate of participation are situated in areas where fertilizer has been demonstrated and used (i.e. by FAO) before the advent of the minimum package project. Such areas are often indentical with areas where a cash crop has been grown for a number of years.

When it comes to the participants it can be noted that the cash buyers form a considerable part, in 1973 they were almost one-third of the total number of participants. Among the credit takers, owner-cultivators formed the overwhelming majority. In 1971 they comprised 82 %, and in 1972

TABLE 12. Cash and credit sales of fertilizer and seeds by the minimum package projects 1971, 1972 and 1973*

	Projects in communally owned areas**	Projects in privately owned areas***
1971		
No. of projects	3	6
No. of cash-sales	70	446
No. of credit-sales	329	1,985
% of credit-sales extended to tenants	–	18 %
1972		
No. of projects	6	12
No. of cash-sales	141	2,388
No. of credit-sales	692	7,521
% of credit-sales extended to tenants	–	14 %
1973		
No. of projects	9	19
No. of cash-sales	500	5,484
No. of credit-sales	4,335	13,328
% of credit sales extended to tenants	–	10-15 % ****

Source: EPID, *Annual Report for 1971/72 Financial Year;* EPID, *Annual Report for 1972/73 Financial Year.*

* The figures refer to the number of registered sales. Since a small number of participants may have registered for more than one sale the actual number of participants is estimated to be somewhat less than the number of sales.

** Eritrea, Tigre, Beghemder and Godjam.

*** Shoa, Wollega, Wollo, Hararge, Kafa, Illubabor, Sidamo, Bale and Gemu Gofa.

**** This percentage is an estimate.

86 %. It is difficult to generalize about this category since it includes both farmers owning and cultivating only a few hectares and landlords cultivating up to 20 hectares and leasing out additional land to tenants. I do not possess systematic information from every project as to the size of the area cultivated and owned by all credit takers. From Shashamane it was reported that a considerable proportion of the credit takers classified as owner-cultivators were actually landlords. On the other hand, almost all credit takers in Kersa minimum package project were reported to own and cultivate less than five hectares.[3] The only conclusion that can be drawn from this is that behind the category of owner-cultivators conceal both small holders who undoubtedly belong to the target group of the projects and landlords who definitely do not belong there.

The tenants form a small part of the participants and, furthermore their proportion is decreasing. In 1971 tenants comprised 18 % of the credit takers and in 1972 14 %. For the year 1973 EPID's annual report does not subdivide the credit takers into tenants and owners, but it is estimated that the proportion of tenants was between 10 % and 15 %.[4] Bearing in mind that tenant holdings are estimated to exceed 50 % of the peasant holdings in the privately owned areas,* it is evident that the most needy group participates only marginally in the credit programme. The general reasons for the low degree of tenant participation has been discussed in the chapters on Bako and Shashamane.

How much have the participants benefitted from the credits? EPID's own evaluation unit has fabricated a formula for the estimation of benefits accrued to farmers as a result of fertilized fields. The formula is constructed in the following way: the incremental crop yields (resulting from the use of fertilizer) are multiplied by crop prices. Costs for inputs (fertilizer) are deducted from this product. According to such a calculation the net benefit for each participating farmer was Eth. $ 75 per hectare in 1972.[5] Although this kind of calculation is basically sound, some comments on it must be made.

First, the prices used in the EPID calculations were the average crop prices in March. March prices were chosen since they were close to the average prices for the whole year. The crux is, however, that the poor peasants sell their produce shortly after harvest, i.e. in December-January, when producer prices are at their lowest. Second, the formula ignores the fact that, when it comes to tenant participants, increased rents generally follow the introduction of the use of fertilizer. In reality, the benefits of the poor peasants were less than EPID's calculations show. Indeed, the extremely low prices in Shashamane-Negele inflicted many participants there with heavy losses.

These comments are not meant to deny that there were farmers who benefitted economically from the credits. But again, the benefits must have been higher among the more wealthy strata of the participants. One should not overlook the possibility that EPID is assisting in the creation of a group of kulaks among the middle and rich peasants, who otherwise would have been locked out from credits, since they are not rich enough to qualify for individual credits from the AID Bank.

The discussion above has tried to show that the majority of the target group of the projects is excluded from participation. Some nuances of this situation will now be given.

* This percentage refers to the mid 1960's.

In the privately owned areas the land holding pattern is not uniform. There are places where land ownerhsip is less concentrated than in Bako and Shashamane. This is the case for example within the project area of Kersa minimum package project in Hararge province and within some extension areas of Wolkite minimum package project in Shoa province. Here almost all peasants are owner-cultivators and absentee landlordism is rare. Thus, tenancy does not create a problem for participation in the credit programme.[6]

On the other hand there are areas where an extreme concentration of land and power in a few hands prevails. In Tulu Bollo minimum package project between the extension areas in Sabata and Ascore there is a large area cultivated by tenants. This area is owned by *Ras* Mesfin, the Governor General of Shoa province. It is well known that the attitude of this landlord towards written lease agreements is entirely negative. None of his tenants can participate in the credit programme of Tulu Bollo minimum package project. Consequently, the project does not try to implement any extension or demonstration activities on *Ras* Mesfin's land, since the project personnel consider these efforts useless. In 1972 the *Ras* had started mechanized farming on a smaller part of his land close to the road.[7]

Within the project area of Wolkite minimum package project a dramatic attempt at mass eviction as a consequence of mechanized farming was undertaken around the turn of the decade. Wolkite is a centre of large scale mechanized farming. The process of mechanization started there in 1969 as a result of fertilizer demonstrations by the earlier mentioned FAO programme. In 1973 some 60 mechanized farms with an average size of 80 hectares had been established.[8]

Teff, wheat and to some extent haricot beans were grown. The commercial farmers are financially strong people from Addis Ababa and local Government officials who have bought up parts of the large areas of Government land which exist around Wolkite. In this area a retired colonel in the Ethiopian army owns 4,000 hectares. He has some 500 tenants cultivating different parts of this land. In 1969 the colonel started mechanized farming. At one point, he made up his mind to expand the process of mechanization rapidly and announced that he was going to terminate the tenancy relationships with all 500 tenants. But the tenants reacted violently. They tried to stone their landlord when he visited a bar i Wolkite. Since that incident the colonel has changed tactics. He gradually mechanizes small areas, thereby evicting only a few tenants at one time. With 240 hectares under tractor cultivation in 1973 this has meant that some 150 tenant families have lost their fields.* These people have not left

* In Wolkite only the colonel's farming enterprise has caused eviction of tenants because the other commercial farmers cultivate land which earlier was idle or grazing land.

the region. Their homes, surrounded by small ensete cultivations are situated in places not suitable for tractor cultivation. While the ensete* provides them with a minimum of food stuffs, a number of them work on the colonel's farm at weeding and harvesting time.

Reforms Prepared at EPID's Headquarters

The weaknesses of the credit programme have been recognized at EPID's headquarters and changes in the regulations have been suggested for the agricultural season 1974.

In order to facilitate participation in the credit programme, it is suggested that tenants not be required to present a written lease agreement with their landlord in order to qualify for fertilizer credits.[9]

At EPID's headquarters it is believed that this regulation has prevented many tenants from participation. Already in 1973 it was the practice however in some projects not to demand written leases.[10] It is assumed that this change will increase tenant participation in the credit programme.

EPID has also suggested that the downpayment should be lowered from the present 25 % to 15 % of the credit value.[11] This suggestion was, however, vetoed by the AID Bank, which follows the principles of the World Bank group.

A new obstacle beyond the control of EPID, threatens the whole credit programme for the year 1974. The price of fertilizer on the world market has increased by 20 % when compared to the 1973 price. According to EPID's calculations, this increase is equivalent to a fertilizer price of Eth. $ 57 per quintal (D.A.P.), i.e. an increase of Eth. $ 19 compared to the price charged in 1973. It is obvious that such a price would prevent most of the poor peasants from participating in the credit programme in 1974. Therefore the Imperial Government decided to give additional subsidies for a certain amount of fertilizer to be distributed by EPID and CADU.[12] With Government subsidies, EPID and CADU will charge Eth. $ 44 for one quintal of fertilizer in 1974. The downpayment for one quintal will then be Eth. $ 11, an increase of Eth. $ 1.5 compared to 1973.

The Ethiopian fertilizer market will, then, offer two categories of fertilizer; unsubsidized fertilizer at a price of Eth. $ 57 per quintal and subsidized fertilizer at a price of Eth. $ 44 per quintal. Limited amounts of the latter category will be available and the intention is that it should be reserved for the target group of CADU and the minimum package projects.

Given this structure of the fertilizer market, competition between com-

* False banana.

mercial farmers and peasants for the subsidized fertilizer is to be foreseen. The commercial farmers will no doubt try to get their share of the fertilizer which is to be distributed through CADU and the minimum package projects. EPID's support to the peasants will have to stand a test during the credit season in 1974.

10.2. Co-operative Promotion

Considering the social structure in southern Ethiopia, it is obvious that there is more to development than the distribution of fertilizer. Indeed, it has been demonstrated above that the use of fertilizer by peasants can act to their own detriment because of a) their powerlessness in controlling prices, and b) the trend towards increasing rents and eviction. EPID is aware of this and its response is the promotion of co-operative societies among the farmers. In EPID's Plan of Operation it was laid down as early as 1971 that co-operative societies were to gradually take over the activities of the minimum package projects. In 1973 the promotion of co-operatives was activated. For this purpose a special 'Section for Co-operative Promotion' was established within EPID. This section includes seven experts operating from the headquarters and a growing number of co-operative organizers (the first 15 were to graduate in mid 1974) who will organize co-operative societies within the particular minimum package projects.[13]

In late 1973 there were a total of 12 registered co-operative societies and 28 pre co-operative societies in various stages of promotion within the areas where minimum package projects operate. Total membership was 10,191. Some of the registered societies had been founded before the advent of the package projects. Among the pre co-operatives some had hardly started any activities at all while others were involved in marketing the production of their members.[14]

The first step in co-operative promotion is that the farmers in a number of villages (100—500 farmers) form a pre co-operative and elect an executive committee. Later, the pre co-operatives within an extension area of a minimum package project merge to form a co-operative society (comprising up to 2,000 members). At this stage the society is registered by the Ministry of National Community Development, i.e. it becomes a juridical person. Since the average minimum package project has 5 extension areas, every minimum package project is supposed to launch five co-operative societies. The aim of the co-operatives is the marketing of produce and procurement of supplies and credit.[15]

The co-operative societies are to perform the following functions:[16]

1) provision of credits for inputs (the societies will act as agents for the AID Bank).
2) distribution of inputs (fertilizer, seeds) to members
3) storage of members' surplus production
4) marketing of members' surplus produciton
5) encouragement of the habit of saving among members
6) education of members in co-operative techniques.

It is suggested that a superstructure of secondary co-operative unions should be organized at a later stage. Every secondary union should comprise 20 primary societies (40,000 members) and take over the functions of storing, transporting, pricing and selling the members' production.[17]

The emphasis is clearly on the commercial functions, the societies are above all to serve as marketing organizations for crops. This is undoubtedly an essential and necessary function, but it is not enough. In order to be able to promote the developmental interests of the poor peasantry beyond mere marketing, the societies must be more than commercial organizations. They must be political organizations.

Such an interpretation is also in accordance with the spirit of the overriding goal for the package projects, as it was originally formulated. But such a political approach to co-operative development is exactly what is lacking in EPID's policy guidelines. A review of the existing power structure within the co-operative societies and EPID's attitude to it will demonstrate this.

The problems concerning the promotion of co-operatives in Bako and Shashamane have been dealt with above. Here, information from the promotion of co-operatives in Wolkite minimum package project will be added. In Wolisso extension area a pre co-operative encompassing somewhat less than 200 members was formed in 1971 on the initiative of the project staff.[18] When it was time to elect the executive committee, the most powerful *balabbat* in Wolisso announced his candidacy for president, and his immense authority secured his way to the presidency. Thus controlling the executive committee, he manipulated the pre co-operative into complete passivity. From 1971 to 1973 the society showed no activity. This is attributed to the fact that the *balabbat*, having heard community development officers speak about the principles of co-operatives, realized that such an organization with active members might pose a challenge to his own—hitherto undisputed—authority among the peasants. Consequently, his tactics were to hinder the activities of the society. Realizing this, the project staff tried at first unsuccessfully to make the *balabbat* resign. In

1973 was the pre co-operative dissolved since it had been completely inactive.

The project staff then started to encourage farmers belonging to the traditional organizations like *mahaber* and *eder* to use their organizations as a base and join together in new pre co-operatives. From the extension areas of Goro and Endeber, in the heart of Gurage land, enthusiasm was reported, but the farmers in Wolisso extension district showed a reluctant attitude because of the failure of the previous attempt.

In Wolkite extension area a pre co-operative comprising some one hundred members was formed in November 1973.[19] All members were small holders including owner-cultivators and tenants. Most of them had participated in the project's credit programme. The tenants in question rented land from small owners, with whom they had close relations. The members planned to have their next harvest sent to Addis Ababa for sale so as to obtain better prices than if they sold individually to merchants in Wolkite. According to the project staff in Wolkite some big merchants and commercial farmers had shown interest in the pre co-operative and wanted to join it. Their reason for this was reported to be that they saw opportunities to use the organizational framework of the co-operative for their own interests. Although the project personnel was well aware of the incompatible interests of the peasants on one hand and the commercial farmers and merchants on the other hand, they did not know how to counteract and they had received no instructions from EPID's headquarters on how to deal with problems of this kind.

The following hypothetical conclusions can be drawn from the experiences from Bako, Shashamane and Wolkite minimum package projects. In areas where feudalism is still 'intact' and commercial farming absent, the local powerholders–the *balabbats* and other large landlords–try to counteract promotion of genuine peasant co-operative societies. This can be done for example by intimidating peasants from joining the societies or by infiltrating the societies' executive committees and thus paralyzing co-operative activity. The hostility towards active co-operative societies stems from the landlords' realization that the societies might become an organizational framework for articulating the economic and political interests of the poor peasants, i.e. activities aimed at undermining the position of the landlords.

In areas where capitalism in the form of large scale mechanized farming has gained terrain, the local powerholders–landlords turned commercial farmers and merchants–view the co-operative societies as organizations which can be used by them in order to create new forms of exploitation of the peasants.

The merchants and commercial farmers can exploit the members as producers by monopolizing the marketing of their crops and they can also exploit them as farm workers. The infiltration by these 'big men' is invited by the legal framework of the Co-operative Societies Proclamation of 1966. This law states that membership in a co-operative society is open to any Ethiopian citizen, be he a farmer or a non-farmer.[20]

10.3. The Reformistic Capacity of EPID

The previous discussion can be summed up like this. The peasants in southern Ethiopia are structurally incapsulated into a larger political system. The rural society in southern Ethiopia is characterized by a marked class structure, the peasants are subordinated to landlords, *balabbats*, merchants and local Government officials. The most clear-cut relation of subordination is that of tenants to landlords. The relations between landlords and tenants are contradictory. The wealth of the former emanates from the toil that the latter perform.

If this analysis is correct, then efforts aimed at improving the living conditions of the poor peasants in general and the tenants in particular must necessarily focus on the structural relationships. Farmers "in the low income brackets" can be helped only through programmes which aim at breaking the exploitative relationships which restrict them to "the low income brackets".

The promotion of co-operative societies *could* be instrumental in such a programme. The condition is that promotion is concentrated to political education of the poor peasants and that landlords and merchants are kept out. Only then could co-operative societies take shape which are fully controlled by the poor peasants forming a solid front against *balabbats*, landlords, commercial farmers and merchants. Such mass organizations are, I believe, a necessary step if the present exploitative structure is to be radically transformed.

Through interviews and discussions with employees at EPID's headquarters, I have been able to verify that the structural problems discussed above are not unknown to EPID. In the Annual Report for 1972/73 it is even mentioned that merchants and landlords often try to sabotage or utilize co-operative societies.[21] But this insight is to a very limited extent reflected in EPID's actual activities. EPID's practical policy for co-operative promotion bears few traces of a structural analysis. Instead, it is recommended that tenants should join the presently existing co-operatives. This recommendation has, so far, not been particularly successful. EPID comments that: "the promotion effort has not yet been successful in convincing tenants that they also may derive benefits from membership".[22] Such policies cannot but suit the interests of the landlords and merchants.

This Janus face of EPID can be understood when one points at two basic facts. One is that there are a number of employees within EPID who are radical sympatizers with the tenants' cause and they can utilize the official, reform oriented attitude of the foreign donor agencies who financially support EPID. The other fact is that EPID is part of the administrative apparatus of the Imperial Government and the Government controls it ultimately. Our final analysis will, then, be focused on the power basis of the Imperial Government and the limits of Imperial reformism.

Notes to Chapter 10

[1] EPID, *EPID Project Areas by Province* (1974).

[2] EPID, *Annual Report for 1972/73 Financial Year*, p. 7.

[3] Information obtained from the project office in Adele (Kersa minimum package project).

[4] Information obtained from EPID's headquarters.

[5] EPID, *Annual Report for 1971/72 Financial Year*, p. 10.

[6] Information obtained from the project offices in Adele and Wolkite. Participation in the credit programme was also considerable. In 1973 Kersa had 652 participants and Wolkite 1,076 participants according to EPID, *Annual Report for 1972/73 Financial Year.*

[7] Information obtained from the project office in Tulu Bollo.

[8] Information obtained from the project office in Wolkite.

[9] Information obtained from EPID's headquarters.

[10] Information obtained from the project office in Wolisso (Wolkite minimum package project) and from EPID's headquarters.

[11] Information obtained from EPID's headquarters.

[12] Ibid.

[13] Ibid.

[14] EPID, *Annual Report for 1972/73 Financial Year*, p. 46.

[15] *Ibid.*, p. 45.

[16] Information obtained from EPID's headquarters.

[17] Bo Wickström, *Increasing Efficiency in the National Grain Marketing System of Ethiopia—A Policy Guideline* (1973), pp. 48 ff.

[18] Information obtained from the project office in Wolisso.

[19] Information obtained from the project office in Wolkite.

[20] Negarit Gazeta, *Proclamation No. 241 of 1966. A Proclamation to Provide for the Formation of Co-operative Societies.*

[21] EPID, *Annual Report for 1972/73 Financial Year*, pp. 47 f.

[22] *Ibid.*, p. 47.

Chapter 11 The Limits of Imperial Reformism—A Case Study and Conclusions

11.1. A Case Study of the Legislation on Agricultural Tenancy Relationships

The analysis of the Imperial Government's propensity to tolerate popular reforms will be approached from a concrete study of one of its major suggested reforms in the field of land tenure; the proposed bill on regulation of agricultural tenancy relationships.

The Ministry of Land Reform and Administration

The Ministry of Land Reform and Administration (hereafter: the Ministry) was established in 1966. Among the Ministry's main tasks are to administer Government land, to map the different land tenure systems in the country and to suggest changes in them which could promote agricultural development. The Ministry has also been given the specific tasks of preparing bills concerning i) registration of land rights, ii) taxation of undeveloped lands and iii) improvement and regulation of landlord-tenant relationships.[1]

The Ministry encompasses many contradictions. This is already evident from its name. The Amharic version of the name contains no reference to reforms, it reads (literally translated to English) 'Ministry of Land Holding and Administration'. The contradictions are further revealed in that one department of the Ministry distributes Imperial land grants and is thus instrumental in the creation of landlords of various magnitudes, while another department of the Ministry deals with measures to restrict the privileges of the landlords.

A detailed study of the working of the Ministry would indeed reveal subtle nuances of the contradictions inherent in the Ethiopian society. Here, the aim is to analyse the content of the results of the Ministry's activities to improve and regulate the tenancy situation.

The Content of the 1971 Draft

In 1968 a first draft of a bill "to provide for the regulation of agricultural tenancy relationships" was prepared in the Ministry and forwarded to the Council of Ministers. It was revised in 1970 and again in 1971. The 1971 draft was submitted to the Chamber of Deputies in the Parliament in

November 1972.[2] This draft was still in early 1974 pending in Parliament. The main paragraphs of the draft bill will now be summarized and commented on.

Before going into details, it is important to note that the bill does not touch on the fundamental issue of land reform—ownership of land. It only specifies the forms of tenancy relationships. The aim of the bill is said to be to encourage landholders and tenants to work in co-operation to increase agricultural productivity and contribute to the future prosperity of Ethiopia.[3] The main provisions are the following: agricultural leases, payment of rent, termination of tenancy relationships, compensation for improvements and pre-emption. It also provides for a machinery of enforcement.

The *lease agreements* can be oral or written. Either party (landholder or tenant) may require that the lease be put into writing. The initiative may also come from the Minister of Land Reform and Administration, who can require that leases in a certain area be put into writing.[4]

The *rent* refers to the payment by the tenant to the landholder for the use of the holding for one year. The rent can be paid in kind or in cash.[5] It may be paid as a fixed amount of agricultural products/money or by the division of the produce of the holding between the landholder and the tenant in proportionate shares. In the case of share-cropping the maximum rent share may not exceed one-third of the gross yield—i.e. the total produce of the holding excluding seed.[6] The rent refers only to payment for the use of land. If the landholder contributes inputs to the holding (bullocks, farm implements, machinery), and/or performs services for the tenant (ploughing, harvesting, threshing, transportation, storage, etc.), the tenant shall give additional payment for these services. The total amount to be paid (rent + other payments) may not exceed one half (50 %) of the gross yield of the holding.[7] If either party so requires, the rent shall be converted to a fixed amount of agricultural products/money.[8] The manner of conversion stipulates that an amount of agricultural products, equivalent to one-third of the gross yield (or a smaller amount in areas where the rent share is traditionally less than 1/3), shall be put aside and recorded as the fixed rent of the holding.[9] Conversion to payment of rent in cash shall be calculated at the current market price of the fixed amount of agricultural products at the market where the tenant usually deals.[10] In cases of widespread disaster the Minister of Land Reform and Administration may order for a reduction in rent or total remittance for one or more rental periods.[11]

Termination of tenancy relationship can be effected on agreement by the parties.[12] In cases where the parties do not agree, termination can be initiated by the landholder as well as by the tenant. The tenant can terminate the relationship by giving one year's notice to the landholder of his intentions. If the tenant can prove that the landholder has failed to comply with any of the provisions of the tenancy proclamation, the tenancy relationship can be terminated without his giving notice to the landholder.[13]

The tenancy relationship can be terminated by the landholder if he can prove that the holding is deteriorating due to tenant negligence (the tenant has abandoned the holding, failed to pay rent, failed to adopt sound farming practices, sub-let the holding, etc.). If, however, the damages can be quickly cured, the tenancy relationship shall not be terminated, provided that the tenant pays damages to the landholder and/or cures the irregularity.[14]

The landholder has furthermore the right to terminate the tenancy relationship due to other causes. The most important of these is for the purpose of introducing large scale commercial farming. The landholder can form a company where he (or his spouse or grown-up child) is the major shareholder. If this company intends to start a large scale commerical farm in the near future, the landholder can terminate the tenancy relationships on the land after giving two years notice to the tenants. If it is considered that the result of such termination would create severe hardship for the tenant, the termination shall be postponed for three more years.[15]

When termination takes place due to this cause, the landholder shall pay disturbance compensation to the tenant, a sum not less than two years rent and not more than five years rent.[16]

If it is discovered that a landholder has evicted tenants in violation of the provisions of the proclamation, it is stipulated that: "the tenant shall be entitled to damages for the losses suffered by him as a result of such eviction . . . and may in addition, be reinstated on the holding where reinstatement is still possible".[17]

Compensation for improvements shall be paid by the landholder to the tenant when the tenancy relationship terminates, whatever the reason for termination.[18] The improvements comprise planting of long term crops (fruit trees, coffee trees), construction of permanent buildings, etc. Compensation shall be equal to the unexhausted value of the improvement at the time of termination.[19]

Pre-emption. If the landholder intends to sell the holding, the tenant(s) shall have the right of pre-emption, i.e. the landholder must in the first case offer the holding at a reasonable price to the tenant cultivating it. If the tenant has not responded to the offer within sixty days, the landholder is free to sell the land to anyone.[20] The right of pre-emption is personal for the tenant and valid only for agricultural purposes.

Implementation and enforcement. The proclamation provides for a three-tier implementation machinery comprising the institutions of Tenancy Officer and Tenancy Committee on the *awradja* level and the institution of Tenancy Tribunal on the provincial level. The Minister of Land Reform and Administration shall appoint a Tenancy Officer, a governmental employee, in each district he sees fit. Concerning application of the different aspects of the proclamation, landholders and tenants can in the first place turn to the Tenancy Officer for guidelines. He is empowered to make investigations of cases of non-compliance with the provisions of the proclamation.[21]

The next link in the implementation machinery is the Tenancy Committee (where the Tenancy Officer acts as secretary). It shall have five or seven members. Landholders and tenants shall be equally represented in the Committee with an owner-cultivator as chairman.[22] Among the duties of the Tenancy Committee are to hear and decide appeals from the decisions of the Tenancy Officer, and to arbitrate disputes betwen landholders and tenants when requested by either party or by the Tenancy Tribunal.[23] The Committee shall make its decisions by majority vote.[24]

The third institution is a special court, the Tenancy Tribunal. It is to be conducted by one judge, experienced in land matters and appointed by the Emperor. The Tribunal is to be supervised by the Ministry of Land Reform and Administration.[25] The Tenancy Tribunal shall function as a court of the first instance dealing exclusively with the regulations of tenancy relationships. It shall hear and decide appeals from the decisions of the Tenancy Committees and its decisions shall have the same effect and be enforced in the same manner as decisions of regular courts.[26] But not even the decisions of the Tenancy Tribunal are final. Discontented landholders and tenants can appeal to the High Court in the province in which the dispute arises. The decision of the High Court is final.[27]

Comments

The proclamation does not rule out share-cropping as a system of rent paying. On the contrary, it is stipulated that the norm shall be one-third to the landholder and two-thirds to the tenant. The conversion to fixed rent has these proportions as its point of departure. Note that this rent only refers to the landholders compensation for agricultural land. If he in addition to the land contributes implements or services, he is entitled to 50 % of the harvest. The provisions thus seem to institutionalize prevailing customs. It is only the most extreme form of exploitation that is forbidden—the practice that the tenant pays 75 % of the harvest as rent.

Concerning termination of tenancy relationships because of landholder's intention to change the use of his land to commercial farming, it is stipulated that the Tenancy Tribunal can, but does not have to, postpone termination by three years, when the two years notice of the landholder has expired. It depends on how the Tenancy Tribunal interprets "severe hardship for the tenant".

Besides, there is the possibility that the landholder and his tenants agree to termination. In such cases the superior bargaining position of the landholder must be taken into account. What formally seems like an agreement on equal terms may in reality be the result of hidden threats and pressure.

The provision that an unlawfully evicted tenant shall have the right to be reinstated on the holding where reinstatement is still possible is also a deceptive formulation. Firstly, the evicted tenants usually leave the region following eviction and are thus out of contact with legal procedures in that region. Even if it is assumed that an evicted tenant should stay in contact with the Tenancy Committee and Tribunal to try to guard his rights, the legal procedure is sure to take a long time. And even if the final decision would be in favour of the tenant the landholder might already have mechanized all his fields. In that case reinstatement is not longer possible. Concerning the provisions of pre-emption it is obvious that they are illusive as long as there exists no credit organization devoted to financing tenants' land purchases.

The provisions on the machinery of enforcement aim at strengthening the tenants' bargaining position. But they are not solid enough. It is unlikely that the Tenancy Officer can work effectively in an otherwise unchanged social structure. If he sides consistently with the tenants he can become the object of undue pressure from the landowners. The usefulness of the Tenancy Committees for the tenants is dependent upon how much influence the tenant representatives can exercise. The highest tier in the new enforcement machinery, the Tenancy Tribunal, is not the final decision-maker. It is possible to appeal to the High Court in the province. It is to be expected that in cases where important economic interests are at stake, the landlords will ultimately ask for justice from the High Courts. It will thus require a time consuming procedure to reach the final decision. And since the ordinary judges as a rule are recruited from the landlord class it is not unlikely that the decisions generally will have a bias in favour of the landlords.

When it finally comes to legal sanctions against violations of the provisions of the proclamation, the penal consequences are vaguely formulated. According to the proclamation, violations against the most important provisions (e.g. illegal eviction) "shall be punishable under the relevant provisions of the Penal Code of 1957".[28] It is not explicitly laid down which these relevant provisions are. The field is open to interpretation. Penal sentences can only be passed by the ordinary courts. The Tenancy Tribunal is stripped of any power in this field. It is laid down in the definition of its material jurisdiction that the Tenancy Tribunal shall "hear and decide any matters or dispute arising out of or concerning alleged violation of any provision of this Proclamation other than the penal provision".[29]

The draft proclamation thus leaves many loopholes which can be used by landlords in order to evade the stipulations, and the enforcement machinery is not given the status necessary for it to provide a solid platform for the articulation of tenants' interests. The proclamation gives the impression of being the outcome of a series of compromises. This impression is confirmed by a comparison with the drafts that preceeded the 1971 draft.

Comparison with Earlier Drafts

The first draft (1968) of the proclamation is on the whole identical with the 1971 draft. Some provisions in the 1968 draft were, however, more radically in favour of the tenants. The most outstanding differences are the following:

i) the 1968 draft states that

> "A landholder who evicts a tenant or otherwise attempts to terminate an agricultural tenancy relationship except in the manner prescribed by this Chapter, as by compelling a tenant to enter into any agreement to terminate such relationship, shall be guilty of an offence under article 554 of the Penal Code".[30]

> Article 554 of the Penal Code stipulates imprisonment or fines for the above offence.[31]

ii) the 1968 draft gives a more active role to the Tenancy Officer. His tasks would have been to

> "bring the substance of the provisions of this Proclamation and of any relevant orders, rules, regulations and decisions made thereunder to the attention of landholders and tenants and promote the understanding thereof by publicity, meetings and other suitable means".[32]

> This provision implies an educational, mobilizing role of the Tenancy Officer.

iii) the 1968 draft stipulates that the highest instance in tenancy disputes shall not be the ordinary provincial court but a special Court of Tenancy Appeals.[33]

These three provisions take a firm stand in favour of giving substance to the improvement of the tenants' legal position. They were thus bound to provoke resistance from the landlords. The resistance was articulated in the Council of Ministers which revised the 1968 draft.[34] The outcome of the revision was the 1970 draft, which was stripped of the three controversial provisions listed above. The 1970 draft even abolished the Tenancy Tribunal and the Court of Tenancy Appeals and made the Tenancy Committee powerless. Otherwise, the content was to a large extent identical with the 1968 draft. As a general comment it can be said that the 1970 draft was vague or silent on points where the 1968 draft was outspoken in favour of the tenants' interests.

The Handling of the 1970 and the 1971 Drafts in the Chamber of Deputies

The 1970 draft was submitted to the Chamber of Deputies in the spring of 1971. It was debated in the spring and autumn of 1971 and met with resistance from a majority of the deputies. No vote was undertaken.[35] The group which had worked out the 1968 draft in the Ministry and had seen its efforts corrupted by the Council of Ministers now embarked on a renewed offensive.[36] The result of the offensive was the 1971 draft which has been dealt with in detail above. The 1971 draft was a compromise in the sense that it was more radical than the 1970 draft but not as radical as the 1968 draft.

The 1971 draft was accepted by the Council of Ministers and submitted to the Chamber of Deputies in the autumn of 1972. It was under debate that autumn and in the spring of 1973. Resistance to the draft among the deputies was still considerable, the debate being at times tense. International pressure was present. The Swedish Government for example had made it understood that Swedish assistance in the field of agricultural development (CADU, EPID) might be discontinued if Parliament had not accepted the draft proclamation before the end of 1972. It was also rumoured that the Emperor wanted the draft proclamation to be accepted.

Nevertheless, 1972 passed, no vote was undertaken and Swedish assistance to CADU and the minimum package projects continued. In the spring of 1973 a showdown was imminent. The group behind the 1971 draft in the Ministry, headed by the Minister of Land Reform and Administration, had during the spring persuaded a majority of the deputies to vote in favour of the draft proclamation. In May 1973 the chairman announced that the Assembly would proceed to vote on the 1971 draft. The vote was, however, prevented because the opponents, who now formed less than 50 % of the deputies, marched out of the Chamber, thus making the assembly formally unable to vote.[37] This incident was followed by intense negotiations the result of which was that the opponents promised to remain at the seats during the next vote and accept a decision in favour of the draft. A few days later the vote was finally to be taken. On the day agreed upon, however, the chairman refused to proceed to the vote, evidently under the instructions of the Emperor personally.[38]

A few days later the term of the Chamber of Deputies ended and Parliament was dissolved.

The case of the tenancy legislation, the most far reaching proposal toward reform up to 1973, shows that the activities and non-activities of the Imperial Government as an institution favoured the interests of the landlord class.

11.2. The Limits of Imperial Reformism: Conclusions

The land and agricultural policies of the Imperial Government, as they are officially laid down in the TFYDP, do not show a clear class bias. Policies aimed at promoting both the interests of the landlords and of the poor peasants are advocated.

But the class bias in the actual execution of the land and agricultural policies is much clearer. The experiences which have been accounted for in this book show that the outcome has been overwhelmingly in favour of the landlords, while tenants have suffered.

The reason for this is that the interests of the landlord class and those of the tenants are contradictory. The commercialization of agricultural production, i.e. the introduction of capitalist relations of production, in areas where the social structure is predominantly feudal has benefitted the landlord class. The landlords are opening up the land, the main means of production, for large scale farming. In this process tenants and poor owner-cultivators are increasingly being evicted or indirectly forced to leave their holdings. A process of polarization and concentration of wealth is taking place.

In this situation, EPID is attempting to promote the interests of the poor peasants. Its credit programme and its co-operative promotion programme have failed to create substantial improvements for the poor peasants in the feudal regions.

If EPID's main goal of popular participation is interpreted in the light of a structural analysis, it would necessarily focus on political mobilization of the poor peasants and the creation of an organizational platform for the articulation of their interests.

But such a programme would logically develop into an outright attack on the present power position of the landlords. This brings us back to the question of the power base of the Imperial Government.

It has earlier been argued that the Imperial Government draws its principal domestic support from the landlord class and the Orthodox Church. In the southern half of Ethiopia local power is vested in people with large landed interests. The support of these landlords is an indispensable basis of power for the Imperial Government.

The Imperial Government cannot be expected to sanction EPID policies which would undermine the strongest local support for its rule, however much individual EPID employees and foreign donors advocate such policies. Under Imperial rule, EPID is bound to abstain from giving substance to the goal of popular participation, interpreted in the spirit of the originally stipulated goals for the package approach to development of peasant agriculture.

It would be wrong to claim that the Imperial Government is opposed to social change. The land and agricultural policies it has sanctioned, have enhanced a pattern of uneven development in the southern provinces. Accumulated wealth for the few and increasing misery for the many is now accelerating as a consequence of the commercialization of agricultural production. By encouraging a capitalist mode of production in the feudal regions, the Imperial Government has rendered new features to the age-old exploitation of the peasants. The southern highlands are being drawn into the modern, dynamic process of underdevelopment.

Notes to Chapter 11

[1] *TFYDP,* pp. 425 f.

[2] *A Proclamation to Provide for the Regulation of Agricultural Tenancy Relationships, Haile Selassie I, Elect of God, Emperor of Ethiopia* (1971), (draft).

[3] *Ibid.,* p. 1.

[4] *Ibid.,* Art. 6.

[5] *Ibid.,* Art. 23.

[6] *Ibid.,* Art. 18.

[7] *Ibid.,* Art. 19.

[8] *Ibid.,* Art. 20.

[9] *Ibid.,* Art. 21.

[10] *Ibid.,* Art. 23.

[11] *Ibid.,* Art. 25.

[12] *Ibid.*, Art. 27.

[13] *Ibid.*, Art. 31.

[14] *Ibid.*, Art. 28.

[15] *Ibid.*, Art. 29.

[16] *Ibid.*, Art. 29.

[17] *Ibid.*, Art. 35.

[18] *Ibid.*, Art. 37.

[19] *Ibid.*, Art. 38.

[20] *Ibid.*, Art. 42.

[21] *Ibid.*, Art. 49.

[22] *Ibid.*, Art. 50.

[23] *Ibid.*, Art. 51.

[24] *Ibid.*, Art. 52.

[25] *Ibid.*, Art. 53.

[26] *Ibid.*, Art. 54.

[27] *Ibid.*, Art. 56.

[28] *Ibid.*, Art. 60.

[29] *Ibid.*, Art. 54.

[30] *A Proclamation to Provide for the Regulation of Agricultural Tenancy Relationships, Conquering Lion of the Tribe of Judah, Haile Selassie, I, Elect of God, Emperor of Ethiopia* (1968), (draft), Art. 61.

[31] *Le Code Pénal de l'Empire d'Ethiopie* (1959), p. 152.

[32] *A Proclamation* . . . (1968), Art. 71.

[33] *Ibid.*, Art. 81–85.

[34] *A Proclamation to Provide for the Regulation of Agricultural Tenancy Relationships, Conquering Lion of the Tribe of Judah, Haile Selassie I, Elect of God, Emperor of Ethiopia* (1970), (draft).

[35] Information obtained from interviews and discussions with people well acquainted with tenancy legislation.

[36] Cf. note 35.

[37] Cf. note 35.

[38] Cf. note 35.

Appendix The Henna Project—an Alternative?

The Henna project is different from the minimum package projects which have been analyzed above. Henna serves as a support organization for a peasant movement which, since the early 1970's is spreading in Ghimbi *awradja* in western Wollega. Henna will be discussed as part of a process of peasant mobilization, which might suggest an alternative to the present gloomy conditions of the Ethiopian peasantry.

1. An Outline of the Social Structure

General Information

The Henna project is active in the western part of Wollega province. This region, between the Didessa river and the Sudan lowlands, is characterized by a hilly, partly mountainous highland at an altitude of 1,600–1,900 m. The climate is temperate in the higher elevations, while some of the valleys are swampy and malaria-infested. The rainy season lasts from late May to late September. The bulk of the population is of Oromo stock, while 'Prenilotic' groups (Bani Shangul) live in the valleys. Agriculture is the main means of sustenance. Maize, teff, sorghum and African millet are the main subsistence crops, coffee being the cash crop.

Christianity is the dominant religion. Besides the Ethiopian Orthodox Church, the Evangelical Church Mekane Yesus has many members and is expanding. Islam is also present.

History

In the sixteenth century pastoral Oromo tribes pushed into Wollega. The earlier inhabitants were thereby pressed down to the river valleys or became absorbed into Oromo society. Even to-day their descendants are found as household servants in Oromo families.

The economy of the western Oromo tribes developed gradually towards agriculture and in this process political authority became increasingly dependent upon the control of land. In the nineteenth century political power had become quite concentrated to a few families. The area which to-day comprises Ghimbi *awradja*, was under the latter part of the nineteenth century under the rule of the Morodda family while nearby Quelem was ruled by Joté Tulu. The incorporation into Menilek's empire followed mostly a non-violent pattern. During the Shoan General Gobana's campaign to Wollega, the Oromo rulers submitted to Imperial supremacy. In return they were provided with regional and local governorships. The land was divided so that 3/4 came under Imperial domain while 1/4 was given to local Oromo rulers. During the twentieth century the land holding system was consolidated through Haile Selassie's land grants.

Land Tenure Pattern in the 1960's

In the mid 1960's there were estimated to be 53,000 agricultural holdings in Ghimbi *awradja*. Of these 58 % were operated by tenants and 42 % by owner-cultivators, 97 % of the holdings comprised 3 hectares or less while 70 % comprised 1 hectare or less.[1] Privately owned land was the most common type of ownership. Within the measured area private land formed 58 %, Government land 18 %, land under *siso* 15 % and Church land 4 %[2]. The largest privately owned holding comprised 19 *gashas* (about 800 hectares).[3] Most landowners lived on their holdings, in a sample of 96 landowners only 12 were absentees.[4]

Among the largest landlords in the area are the descendants of the great Oromo rulers. Challia *woreda* is virtually under one-man ownership. In Quelem *awradja* the descendants of the ninteteenth century ruler Joté Tulu own large areas and numerous villages.[5] Within the Henna project area there are approximately 30—50 large landlords holding 10—30 *gashas* each and having numerous tenants. But there are also a great number of small owners, cultivating a few hectares and leasing out another few hectares to close relatives.[6] Thus, it is a matter of terminology whether farmers cultivating fields of close relatives should be classified as tenants or as farmers within a family ownership system.[7]

Paying rent lies at the core of the tenants' obligations. If the tenant is a share-cropper, the share varies between 1/4 and 1/2 of the harvest. Payment in cash is the most common form of payment.[8] Coffee is cultivated on a great many small holdings and provides as a cash crop the peasants with money.[9] Apart from paying rent, it is common that tenants perform all sorts of agricultural work on the fields which are utilized for the landlord's household. In general, the obligations of tenants cultivating the land of relatives are less demanding.[10]

Mechanized agricultural production has not reached Ghimbi. Communications are poor. The main road ends in the town of Ghimbi, from there a fourwheel drive vehicle is necessary. The terrain is hilly and discourages cultivation by tractor.

The Evangelical Church Mekane Yesus

West of the Didessa valley and especially in Ghimbi *awradja*, the Evangelical Church Mekane Yesus is a religious institution of importance. Originally an evangelical missionary station, the Church has gained foothold among the peasants and was registered as an independent Ethiopian church in 1958. In 1972 Mekane Yesus had 88,116 members and a wide group of supporters in Western Wollega. Most of the members and the church activities are found in Ghimbi *awradja*.[11]

The Church has considerable grass root support from the peasant population. The reasons for this are said to be historical. At the time of the incorporation of Wollega into the Ethiopian empire, the Ethiopian Orthodox Church, which the Oromo rulers joined, was imposed on the people and a church tax was introduced. The Ethiopian Orthodox Church came to be identified with the landlords and the political authorities.[12] In contrast to this, the missionaries who arrived early in the twentieth century (Eritreans and Oromos and later Swedes and Germans) settled among the peasants, learned their language, and built schools and hospitals. Soon the Church was anchored in the villages and priests and congregation leaders were recruited from the local population. When restrictions on evangelical work were lifted in the 1950's, the membership of Mekane Yesus expanded rapidly. Church activities have emphasized social work. In 1962 literacy campaigns started and have contributed to a significant increase in the rate of literacy among the peasant population. According to a survey of the Henna area 26 % of the members of the surveyed households were literate in 1972, when compared to an estimated 5 % for Wollega as a whole.[13]

The significance of the Church for the following discussion lies in the fact that it constitutes a mass organization with a certain degree of democratic traditions (elections to and discussions within congregations and parishes). A considerable part of the peasants involved in the peasant movement in Ghimbi *awradja* are members of the Evangelical Church Mekane Yesus.

2. The Peasant Movement

A broad outline of the events that led to what in 1973 definitely could be characterized as a mass movement among peasants in parts of Ghimbi *awradja*, will now be made.

The Spark

In 1971 a few Community Development workers from the Ministry of

National Community Development were stationed in Nedjo.[14] These CD workers were devoted to organizing the peasantry and played a significant role in the early stages of the movement. They first approached the large landlords telling them that they had been sent out by the Government to teach the tenants better agricultural practises, from which the landlords would benefit in terms of increased rents. This introduction gained the landlords' approval of the CD workers' suggestion to organize meetings with the tenants. A first massmeeting was held at Nedjo airstrip on a Sunday, when the landlords had ordered their tenants to be present. Approximately 5,000 peasants gathered on the airstrip where the CD workers, using a megaphone, announced that they had been sent in order to help the small farmers with their daily agricultural problems. The peasants responded by inviting the CD workers to their villages for further discussions.

During all of 1971, CD workers visited the small villages in the Nedjo-Jarso-Bodji area one by one. During the visits the discussions were broadened from day to day problems to focus on the contradictions within the local society. More concretely, CD workers stressed the fact that the peasants spent yearly considerable sums of money on court cases disputing landmatteres. The CD workers claimed that this praxis only served to enrich the lawyers and judges who recieved fees and bribes while the peasants became more impoverished. This state of affairs could be changed, the CD workers further claimed, only if the peasants within a village could agree to elect a committee from their own members, which would serve as a first dispute-settling instance. By referring disputes to this committee, the parties involved would save money (court fees, bribes, living expenses while in the court town) and time (court cases are time-consuming events when the farmers involved must stay away from their holdings). If the elected committee failed, the parties could then proceed to the ordinary courts. This reasoning convinced the peasants, and a number of dispute-settling committees were set up.

Since 1971 lawyers and judges in the *woreda* courts are reported to have complained over decreased incomes, which is an indication of the success of the committees.

Soon the CD workers suggested that the village members should join in a pre co-operative society and elect an executive committee to complement the dispute-settling committee. The function of the pre co-operative would be to institutionalize co-operation among the villagers. This suggestion was also accepted in a number of villages and the first activity of the pre co-operatives was to build a house, where the villagers could gather to discuss common problems.

During the same period, CD workers had grown teff on a demonstration plot using fertilizer. The result was that the plot yielded almost five-times a normal harvest. This yield became widely known and is reported to have had a striking effect on the peasants.

As a consequence of over-cultivation the harvests in Ghimbi *awradja* had continually diminished during the 1960's. Many peasants despaired and thought of moving down to the lowlands. In such a situation the fertilizer arose general enthusiasm. The information was spread through the villages and many peasants decided to try fertilizer for the agricultural year 1972.

Henna Project Activities

It was at this stage that the Evangelical Church Mekane Yesus became involved. The Church had for some time planned a rural development programme to be executed in Ghimbi *awradja*, where a large number of peasants were churchmembers. The aim was stated to be promotion of a co-operative organization among the peasantry, and an extension service in the field of crop and animal husbandry.[15] The project centre was built in Henna and the project is thus called the Henna project. The Church contacted EPID concerning supplies of fertilizer for Henna. An agreement was reached, according to which EPID acts as the supplier of fertilizer from 1972 onward.[16] The Henna project established three extension districts for the agricultural season 1972. These were in Nedjo, Bodji and Jarso. Among the first activities were the sales of fertilizer.[17]

When Henna started its activities, it rode on a wave of enthusiasm for fertilizer created by the CD workers' early experiments. During the summer of 1972, about 700 peasants, most of which were tenants, bought fertilizer from the project*.[18]

The increase in yields due to fertilizer further encouraged the villagers to form co-operatives. In late 1972 more than 100 incipient co-operatives, with a total membership of approximately 11,000 had been formed.[19] This figure corresponds to about one-fifth of the peasant households in all Ghimbi *awradja*.

In the villages which established co-operatives, discussions continued and the CD workers, from the summer of 1972 onward together with the Henna project's staff, were here instrumental in articulating the problems which had to be faced.[20] Two spheres of activity for the co-operatives were specified. The first was to improve communications. The road network

* No fertilizer was given on credit since it was known that the peasants had cash incomes from the sale of coffee. The project staff argued that credits would create problems with repayments and thus conflicts between the project and the peasants.

was extremely deficient. The villages needed roads for facilitating the transport of goods to market, ambulance service, the safe conduct of children to school, funeral processions etc. Since the Government did not provide roads, some co-operatives decided to start building roads by themselves. This denoted an enlarged scope of the co-operation since road building necessitates the joint action of several villages.

The second activity was co-operation in agriculture during the high seasons of ploughing and harvesting. Here co-operation could build upon old traditions, amongst which the most important is *dabo*. *Dabo* means that a number of neighbouring peasants together perform a specific agricultural task (for example ploughing). In this way the fields of one member at a time are worked on collectively. The harvest from the fields belongs to the individual members. Agricultural co-operation during 1972 consisted simply of institutionalizing *dabo* within the pre co-operatives.

That which gives the peasant movement in Ghimbi *awradja* its profile is, however, that it articulated the contradictions between the tenants and the large landlords.[21] The CD workers from the very beginning in 1971 agitated against the exorbitant exploitation of the tenants by the landlords* in the form of day-works, the custom of presenting gifts to landlords etc. The result of the discussion in pre co-operatives on this point was that starting in 1972 a growing number of tenants organized in pre co-operatives refused to give other tributes to the landlords than rent (in cash or share of crop). As a consequence the landlords had to hire agricultural workers to cultivate the fields reserved for their own households. This made the conflict overt, and focused the landlords' attention on the most articulated CD worker who had been instrumental in the formation of many pre co-operatives. When this CD worker was involved in an attempt to close a police station, the landlords saw their opportunity to take action. The CD worker was accused in court of disobeying Government officials. He was sentenced to leave Wollega province and not to return.

But the peasant movement did not die out because of the removal of one of its initiators. The movement had since mid 1972 received increasing support from Mekane Yesus' Henna project, which had been actively encouraging the villagers to continue their co-operative efforts.

A few figures can illustrate the growth of the movement in 1973. A total of 2,305 peasants bought fertilizer in 1973 from the Henna project. Of the buyers 42 % were tenants, 90 % bought one bag (= 1/2 quintal) of fertilizer or less. Some 30 landlords participated in the programme, buying 10–12 bags each.[22] At the end of 1973 the number of co-operative

* The following discussion is in most cases not valid for the relations between tenants and owners within the above mentioned "family-ownership" system of land tenure.

members was estimated to be close to 15,000. The number of pre co-operatives exceeded 100 but was not exactly known. Information was continually coming to the project office that additional villages had decided to form pre co-operatives. Also villages outside the project area turned to the project staff for assistance in the formation of pre co-operatives.[23]

The main activities of the pre co-operatives (road building and agricultural co-operation) progressed during 1973. In the autumn of 1973 two road-building projects were underway, one from Bodji-Bila to Ghimbi through densely populated areas (about 30 km), the other from Bodji-Bila to Dongoro (about 15 km). Both roads have been initiated by the villagers who live in the immediate area and who have joined in co-operatives. The road Bodji-Bila–Dongoro has been undertaken by three co-operatives, each of which is responsible for a particular stretch of road. The co-operatives have, however, also worked on some sections together. When I visited the two road-building projects in November 1973, work was undertaken one day a week and this continued until the harvest started in December. At the time of my visit up to 300 men in each project were engaged in digging. The work proceeded at a forced tempo. Governmental orders did not spur the work, but popular enthusiasm did. The work was coordinated by a priest form Mekane Yesus. After the work day the men gathered to decide which day to continue the work. The priest made a short speech on the necessity of continued efforts for the wellbeing of the villagers. A prayer ended the meeting.

In the field of agricultural co-operation some pre co-operatives proceeded a step beyond the *dabo* in 1973. One pre co-operative had contracted a hillock (appr. 10 hectares) close to the Henna project center from a landlord on cash basis. The agricultural work on that hill was performed by the members together. At ploughing time more than one hundred oxen were working the field at the same time. The harvest was also to be performed collectively and the crop to be shared among the members. This is the only organized *production* co-operative that I have heard of among peasants in Ethiopia.

These impressionistic bits of information may suffice. It now remains to discuss the options for the future of the peasant movement in Ghimbi *awradja*.

Present and Future Problems

According to my information the movement started (the dispute settling committees)as a response to exploitation by the authorities. It gained terrain with the introduction of fertilizer, which in this case meant a considerable improvement in the peasants' food provisions. Encouraged by the improvement in agriculture the movement expanded in size and activities (the road-building) and articulated the structural contradictions implied in the tenancy conditions (the refusal to work for the landlords). This blow at the landlords' interests provoked action from the authorities (the accusation of the CD worker). Attempts at repression were instigated: when 3 secretaries of pre co-operatives visited the town of Ghimbi in 1972 they were arrested and brought to the provincial prison in Nakamte, where they were kept for several months accused of "disobedience of Government officials". They were later on released.

It is quite logical that the large landlords and the local authorities (*awradja* and *woreda* governors) take a hostile attitude toward the peasant movement. And it is also quite logical that the peasant movement, in the promotion of its members' welfare, is becoming more articulate in its position against the large landlords.

It would be futile to try to predict the future development of the confrontations in detail. But some general outlines can be discussed.

Consider first the peasants. It has been shown many times in history that although the peasants represent a formidable physical force, peasant movements have failed due to deficient or inaccurate organization. In Ghimbi *awradja*, the catalyst for the movement was provided from outside. Since 1972 the foremost institutionalized support for the movement has been the Evangelical Church Mekane Yesus in the form of the Henna rural development project. The project activities are centred around agricultural extension and the promotion of co-operatives. Additional activities are courses in home economics and animal husbandry. The principles for approaching peasants with these activities are stated thus:

> "The main effort of the project is based on an extension program. The extension approach has been described as a system of out-of-school education for rural people. Instead of calling the farmers to a center for education, the instructors are going out to the farmers instructing them on their own fields.
>
> The extension workers should never impose their ideas on the people. Under positive influence the farmers themselves must always be the responsible and active partner in any decision and action. The extension workers, often called the agents, should live among the people, work with them, observe their problems, and then encourage the people to solve the problems individually or cooperatively.
>
> The extension service should include the whole family, father, mother and children. The secondary effects, sociological and cultural, have always to be considered."[24]

Considering the most immediate problem of the peasants, food production, it is obvious that the project can offer valuable assistance. The hidden food shortage among the peasants is shown by the fact that the increase in teff production (teff is the staple food of the peasants) which occurred as a result of fertilization, showed up in the local markets first in 1973 and then only on a moderate scale. The peasants, then, consumed the whole 1972 increase and much of the 1973 increase. Only in 1974 will teff come out on the markets in large quantities. This development is foreseen by the project staff and the activities in agricultural extension will be focused on the improvement of crops other than teff, preferably food crops such as pulses and cash crops like oil seeds and coffee. Nevertheless, the widespread and growing enthusiasm for fertilizer will, during the coming few years (from 1974 on), create a surplus of crops for the market. Here one important point arises. As has been pointed out earlier an increase in production without a concomitant organization of the producers will only lead to a decline in market prices, i.e. profits for the merchants and losses for the peasants. This situation will arise with the 1974 harvest. Therefore the project will give priority to the organization of marketing channels for the pre co-operatives. This includes contracting trucks for transportation, storing facilities, opening up contacts with buyers in Addis Ababa etc.—tasks which the peasants at this stage will not be able to do by themselves. The efforts to establish a peasant based marketing apparatus will probably lead to a confrontation with the local merchants, who can counteract for example by increasing the price of items like salt, which the peasants are dependent on them for. The problems are bound to be many and it is futile to try to suggest the outcome.

In the field of marketing agricultural production, the enemies of the peasants are the merchants. If the peasants are firmly supported by the Henna project, the merchants should not prove to be an insurmountable obstacle.*

The problem is more difficult when it comes to the political aspects of the peasant movement. Here many options are open. *If* the movement limits itself to agricultural production and to marketing, essential political issues will not necessarily be raised. But *if* the peasants pursue their interests further—which is indicated by their refusal to give free labour to the landlords—a confrontation with the local powerholders is inevitable, due to the prevailing structure of the society. Many issues can be raised that immediately confront the powerholders. One issue could be that the tenants

* Information from the Henna project which reached me at the time of printing this book, indicates that the merchants in Ghimbi *awradja* are trying to get membership in the co-operatives and thus influence them from inside.

demand ownership of the holdings they cultivate. Another could be the demand that the local governors should actively work for increased public services which benefit the whole population of the *awradja* or, still more, to demand participation in decision making and the allocation of tax revenues.

Provided with an organizational framework (co-operatives, the Henna project, the Mekane Yesus Church) and inspired to act for social justice, the peasant movement will constitute a formidable social force, against which thc numerically few large landlords, merchants and governors are nothing—as individuals. But again, the outcome of such a confrontation depends to a great extent upon the support which the powerholders can mobilize. The local authorities might ask the Imperial Government to intervene. The Imperial Government has never allowed peasants to organize themselves for political ends. Some attempts (i.e. the Mecha-Tulama movement in the 1960's) have ended with death sentences for the leaders.[25] It is quite logical that, as long as the power basis of the Imperial Government is the landlords, it can tolerate no mass peasant organization to work openly politically.

But the movement in Ghimbi is different from the Mecha-Tulama movement. In Ghimbi the peasants are responding to the official Government policy of encouraging small scale farming with technical improvements. As long as that is the foremost activity of the movement the authorities (local and national) can not openly go against it. It is only if political demands are articulated that the Government will react. Many repressive actions could be undertaken against the peasants. They range from refusal to register the pre co-operatives (denying them official juridical status), refusal to provide fertilizer to the area (the Henna project gets fertilizer through EPID), arresting peasant leaders for 'disobedience', or ultimately sending units of the Ethiopian army against the peasants. How far the Imperial Government will go if confronted with a politically conscious and geographically expanding peasant movement is difficult to predict. But no action should be considered out of question.

Then there is the Evangelical Church Mekane Yesus. Will it continue to give firm support to the peasant movement even if the movement develops in a radical direction?

And as for the peasants themselves, they have hitherto been referred to as a homogenous mass. Actually there are many different categories of peasants active in the movement. First there are the tenants who form the poorest section, whether they are tenants to relatives or to large landlords. Second there are the owner-cultivators who own no more than they cultivate through their own family labour. These two categories form the majority of the peasants, but there is a third: the peasants who themselves cultivate some hectares but own more and lease out their additional land. The upper stratum of this category can easily be integrated with the landlords.

The owner cultivators which lease out additional land to tenants often have prominent positions in the co-operative movement. They are presidents of pre co-operatives and the like. It can be argued that this group has more to loose from a confrontation with the authorities and therefore might refrain from giving priority to political issues within the co-operatives. When the marketing of crops becomes a reality for the co-operatives, misappropriation of funds will be an ever present temptation for the co-operative leadership. This could also cause splits.

There are, thus, different strata with different interests within the emerging peasant movement. And, so, the risk of disintegration of the movement is imminent when the movement meets acute crises in the future.

Still, the peasant movement in Ghimbi indicates an alternative road to development for Ethiopia, that of releasing the productive forces through mobilization of the popular forces for popular interests.

Notes to Appendix

[1] *Survey of Wollega*, pp. 26 f., 29.

[2] *Report on Land Tenure Survey of Welega Province*, p. 6.

[3] *Ibid.*, p. 13.

[4] *Ibid.*, p. 16.

[5] Information obtained from interviews and discussions with people well acquainted with the Ghimbi area.

[6] Information obtained from the project office in Henna.

[7] Eigil Sanna, *Henna Project, Report on Fertilizer Sale* (1973), p. 3.

[8] *Survey of Wollega*, p. 30.

[9] Hunduma Gudetta, *A Regional Study of Aira* (1969), p. 52.

[10] Cf. note 5.

[11] Information obtained from the Evangelical Church Mekane Yesus, Bodji District.

[12] Negaso Gidada with Donal Crummey, *op.cit.*

[13] Information obtained from the Evangelical Church Mekane Yesus, Bodji district.

[14] Cf. note 5.

[15] Evangelical Church Mekane Yesus, *Rural Development Programme, Western Synod.*

[16] EPID, *Annual Report for 1972/73 Financial Year*, p. 6.

[17] Information obtained from the project office in Henna.

[18] Henna Project, *Fertilizer and Variety Trials 1972* (1973).

[19] Information obtained from the project office in Henna.

[20] Ibid.

[21] Cf. note 5.

[22] Sanna, *op.cit.*, pp. 3 f.

[23] Information obtained from the project office in Henna.

[24] Evangelical Church Mekane Yesus, *Henna Pilot Project for Rural Development in Ethiopia, Annual Progress Report 1973* (1974), p. 2.

[25] Clapham, *op. cit.*, pp. 81, 198.

Bibliography

Books, articles, papers, reports.

Aberra Moltot, The Need for Public Participation in the Executions of Land Reform Laws, *Seminar Proceedings on Agrarian Reform* (Addis Ababa 1970).

Abir, M., Ethiopia: *The Era of the Princes. The Challenge of Islam and the Re-unification of the Christian Empire 1769—1855* (London 1968).

Alavi, H., Peasant Classes and Primordial Loyalties, *The Journal of Peasant Studies*, Vol. 1, No. 1, 1973.

de Almeida, M., The History of High Ethiopia or Abassia, *Some Records of Ethiopia 1593—1646*.

Alvares, F., *The Prester John of the Indies*, ed. by C.F. Beckingham and G.W.B. Huntingford, Vol. I (Glasgow 1961).

Ambaye Zekarias, *Land Tenure in Eritrea (Ethiopia)* (Addis Ababa 1966).

Amin, S., Underdevelopment and Dependence in Black Africa, *Journal of Peace Research* No. 2, 1972.

Amin, S., *Le Développement Inégal. Essai sur les Formations Sociales du Capitalisme Périphérique* (Paris 1973).

Asfaw Edessa, *A Micro-Geographical Study of Mankata* (Addis Ababa 1970) (mimeo).

Asmarom Legesse, *Gada, Three Approaches to the Study of African Society* (New York 1973).

Bahrey, History of the Galla, *Some Records of Ethiopia 1593—1646*.

Bairu Tafla, Ras Gobäna Daci Abba Tegu 1821—1889, *Journal of Ethiopian Studies*, Vol. V, No. 2, 1967.

Bairu Tafla, Marriage as a Political Device: An Appraisal of a Socio-Political Aspect of the Menilek Period 1889—1916, *Journal of Ethiopian Studies*, Vol. X, No. 1, 1972.

Bartels, L., Studies of the Galla in Wällägä, *Journal of Ethiopian Studies*, Vol. VIII, No. 1, 1970.

Bauer, D.F., *Land, Leadership and Legitimacy among the Inderta Tigray of Ethiopia* (Michigan 1972) (Unpublished Ph.D. dissertation).

Belayneh Negatu, *The Use of Development Conditions on Government Land Grants in Ethiopia: Practise and Prospects* (Addis Ababa 1969) (mimeo).

Bergman, G. and Lindqvist, H., *Credit Situation in Chilalo Awraja* (Minor Research Tasks at CADU No. 3, 1969).

Berhanu Gebre Mariam, Shashamane and the Agricultural Development Project, *HSIU and USAID Summer Interns* (Addis Ababa 1970).

Berhanu Gebre Mariam, *Adoption of Innovation at Bako Minimum Package Project* (Addis Ababa 1973) (mimeo.).

Bondestam, L., *People and Capitalism in the Awash Valley-Ethiopia* (Dakar 1973) (mimeo.).

Bondestam, L., *Some Notes on African Statistics — Collection, Reliability and Interpretation* (Uppsala 1973).

Bondestam, L., *Den dömda dalen — om kapitalism och människor i nordöstra Etiopien* (Lund 1974).

Borderon, A., *Introduction to the Southern Region's Agricultural Development Project* (Addis Ababa 1973) (mimeo.).

Carlsson, T., - Carlsgårdh, A., *Road Construction Project, Bako-Jere, Final Report* (Addis Ababa 1973) (mimeo.).

Clapham, C., *Haile-Selassie's Government* (London 1969).

Clark, G. and Piggott, S., *Prehistoric Societies* (London 1970).

Le Code Penal de l'Empire d'Éthiopie (Paris 1959).

Debelian, L., *The Impact of Agricultural Mechanization on the Traditional Society in Ethiopia* (Dakar 1973) (mimeo.).

Desta Asfaw, Tesfaye Wolde Mariam, Wondwossen Hailu and Zemedhun Wolde Gabriel, *Nomadic Areas* (Addis Ababa 1973) (mimeo.).

Deutsche Aksum-Expedition Band I, II, III, IV (Berlin 1913).

Dillnessa Zewdie, *Land Use Study. Ten Selected Farmers in Yebrage Hawariat* (Addis Ababa 1971) (mimeo.).

Doresse, J., *Histoire Sommaire de la Corne Orientale de l'Afrique* (Haarlem 1971).

Dunning, H., Land Reform in Ethiopia; A Case Study in Non-Development, *UCLA Law Review*, Vol. 18, No. 2, 1970.

Ellis, G., *Man or Machine Beast or Burden. A Case Study of the Economics of Agricultural Mechanization in Ada District, Ethiopia* (Michigan 1972) (Unpublished Ph.D. dissertation).

Evangelical Church Mekane Yesus, Rural Development Programme, Western Synod.

Evangelical Church Mekane Yesus, Henna Pilot Project for Rural Development in Ethiopia, Annual Progress Report 1973 (Henna 1974) (mimeo.).

FAO, *Freedom From Hunger Campaign Fertilizer Programme, Near East and North Africa* (Rome 1968).

Final Report on the Appraisal of CADU and EPID by the Evaluation Team appointed by the Imperial Ethiopian Government and the Swedish International Development Authority (Stockholm 1974) (mimeo.).

Gabra Selase, *Chronique du Régne de Ménélik II Roi des Rois d'Éthiopie*, Vol. 1, 2 (Paris 1930–32).

Galtung, J., A Structural Theory of Imperialism, *Journal of Peace Research*, No. 2, 1971.

Gizachew Adamu, *A Historical Survey of Taxation in Godjam 1901–1969* (Addis Ababa 1971) (mimeo.).

The Glorious Victories of Amda Seyon, ed. by G.W.B. Huntingford (Oxford 1965).

Greenfield, R., Ethiopia, *A New Political History* (London and New York 1965).

Haberland, E., *Galla Süd-Äthiopiens* (Stuttgart 1963).

Haberland, E., *Untersuchungen zum Äthiopischen Königtum* (Wiesbaden 1965).

Habtemariam Seyoum, *Military Organization and Armament Acquisition of Menilek II* (Addis Ababa 1967) (mimeo.).

Haile Menkerios, The Present System of Land Tenure in Ethiopia, *Challenge*, Vol. X, No. 2, 1970.

Haile Selassie I, *Selected Speeches* (Addis Ababa 1967).

Haile Wolde, *Problems of Land Tenure in the Implementation of CADU Project* (Addis Ababa 1973) (mimeo.).

Henna Project, Fertilizer and Variety Trials 1972 (Henna 1973) (mimeo.).

Henock Kifle, *An Analysis of the Cadu Credit Programme 1968/69 – 1970/71 and its Impact on Income Distribution* (CADU Publication No. 66, 1971).

Henock Kifle, *Investigations on Mechanized Farming and its Effects on Peasant Agriculture* (CADU Publication No. 74, 1972).

Hess, R.L., *Ethiopia, The Modernization of Autocracy* (Ithaca 1970).

Hoben, A., Land Tenure and Social Mobility among the Damot Amhara, *Proceedings of the Third International Conference of Ethiopian Studies*, Vol. III, (Addis Ababa 1970).

Hodson, A.W., *Seven Years in Southern Abyssinia* (London 1927).
Huffnagel, H.P., *Agriculture in Ethiopia* (Rome 1961).
Hultin, J., *The Galla Expansion Reconsidered* (1974) (mimeo.).
Hunduma Gudetta, *A Regional Study of Aira* (Addis Ababa 1969) (mimeo.).
Huntingford, G.W.B., *The Galla of Ethiopia, the Kingdoms of Kafa and Janjero* (London 1955).
Huntingford, G.W.B., *The Land Charters of Northern Ethiopia* (Addis Ababa 1965).
IBRD, *The World Bank Group in Ethiopia* (Addis Ababa 1967).
IBRD, *The World Bank Group in Ethiopia* (Addis Ababa 1973).
Johansson, A., Gobena överger aldrig sin jord, *Dagens Nyheter*, June 5th, 1974.
Jones, A.H.M., and Monroe, E., *A History of Ethiopia* (Oxford 1970).
Kebede Tessemma, *Land Use Study in Loke* (Addis Ababa 1972) (mimeo.).
Kelber, E., – Awole Mela, *Agricultural Survey of Bako Area* (Addis Ababa 1970).
Kuls, W., *Bevölkerung, Siedlung und Landwirtschaft im Hochland von Godjam* (Frankfurt am Main 1963).
Lawrence, J.C.D. and Mann, H.S., F.A.O. Land Policy Project (Ethiopia), *Ethiopia Observer*, Vol. IX, No. 4, 1966.
Legesse Gebeyehu, *Conquest of the Kingdom of Kaffa* (Addis Ababa 1971) (mimeo.).
Levine, D.N., *Wax & Gold. Tradition and Innovation in Ethiopian Culture* (Chicago 1972).
Lewis, H.S., *A Galla Monarchy:Jimma Abba Jifar, Ethiopia 1830–1932* (Madison and Milwaukee 1965).
Lewis, I.M., *Peoples of the Horn of Africa. Somali, Afar and Saho* (London 1955).
Lexander, A., *The Changing Rural Society in Arussi Land: Some Findings from a Field Study 1966–67* (CADU Publication No. 7, 1968).
Littman, E., Reisebericht der Expedition, Topographie und Geschichte Aksums, *Deutsche Aksum-Expedition*, Band I.
Lundberg, J.O., *Haro- ett stycke u-landsverklighet, ett samhällsreportage om utveckling och underutveckling i Etiopien* (forthcoming).
Lundström, K.J., *Communal-Individual Land Ownership System in Northern Ethiopia* (Uppsala 1974) (mimeo.).
Mahteme Sellassie, The Land System of Ethiopia, *Ethiopia Observer*, Vol. 1, No. 9, 1957.
Marcus, H.G., Imperialism and Expansionism in Ethiopia from 1865 to 1900, *Colonialism in Africa 1870–1960,*Vol. 1, ed. by L.H. Gann and P. Duignan, (Cambridge 1969).
Meheretab Tekie, *A Case Study of a Small-Scale Commercial Farm* (Addis Ababa 1969) (mimeo.).
Mesfin Wolde Mariam, *An Atlas of Ethiopia* (Addis Ababa 1970).
Mesfin Wolde Mariam, *An Introductory Geography of Ethiopia* (Addis Ababa 1972).
Michael Beyene, *An Analysis of CADU Credit Programme 1971/72 – 1972/73* (CADU Publication No. 92, 1973).
Mintz, S., A Note on the Definition of Peasantries, *The Journal of Peasant Studies* Vol. 1, No. 1, 1973.
Mulugeta Tessemma, *Land Use Study in Lege Batu* (Addis Ababa 1971) (mimeo.).
Murdock, G.P., *Africa. Its Peoples and their Culture History* (New York 1959).
Negaso Gidada with Donald Crummey, The Introduction and Expansion of Orthodox Christianity in Qelem Awraja Western Wällägä from about 1886 to 1941, *Journal of Ethiopian Studies* Vol. X, No. 1, 1972.
Negatu Lakew, State and Land Grant System in Ethiopia, *Seminar Proceedings on Agrarian Reform* (Addis Ababa 1970).

Nekby, B., *CADU An Ethiopian Experiment in Developing Peasant Farming* (Lund 1971).

Orent, A., Refocusing on the History of Kafa prior to 1897, A Discussion of Political Processes, *African Historical Studies*, Vol. III, No. 2, 1970.

Pankhurst, R., *The Great Ethiopian Famine of 1888–1892: A New Assessment (Addis Ababa 1964) (mimeo.).*

Pankhurst, R., *State and Land in Ethiopian History* (Addis Ababa 1966).

Pankhurst, R., *An Introduction to the History of the Ethiopian Army* (Addis Ababa 1967).

Pankhurst, R., *Economic History of Ethiopia 1800–1935* (Addis Ababa 1968).

Pankhurst, R., A Page of Ethiopian History: Italian Settlement Plans during the Fascist Occupation of 1936–41, *Ethiopia Observer* Vol. XIII, No. 2, 1970.

Papers in African Prehistory,ed. by J.D. Fage & R.A. Oliver (London 1970).

Pausewang, S., Peasant Society and Development in Ethiopia, *Sociologia Ruralis*, Vol. XIII, No. 2, 1973.

Perham, M., *The Government of Ethiopia* (London 1948).

Preliminary Report of the Appraisal Team on CADU (1969) (mimeo).

Some Records of Ethiopia 1593–1646, ed. by C.F. Beckingham and G.W.B. Huntingford (London 1954).

Redden, K., Aberra Jembere and Nabiyelul Kifle, *The Law Making Process in Ethiopia* (Addis Ababa 1966).

Report on a Survey of Begi and Gidami Woreda (1973) (mimeo.).

Rubenson, S., *Wichale XVII. The Attempt to Establish a Protectorate over Ethiopia* (Addis Ababa 1964).

Rubenson, S., *King of Kings Tewodros of Ethiopia* (Addis Ababa 1966).

Sanna, E., *Henna Project, Report on Fertilizer Sale* (Henna 1973) (mimeo.).

Schwab, P., *Decision-Making in Ethiopia, A Study of the Political Process* (London 1972).

Sergew Hable Sellassie, *Ancient and Medieval Ethiopian History to 1270* (Addis Ababa 1972).

Shiferaw Melesew, *A Land Use Study of Botche* (Addis Ababa 1971) (mimeo.).

Sik, E., *The History of Black Africa* Vol. 1. (Budapest 1966).

Simoons, F.J., Some Questions on the Economic Prehistory of Ethiopia, *Papers in African Prehistory*.

The Somali Peninsula. A New Light on Imperial Motives (Mogadishu 1962).

Stitz, V., The Amhara Resettlement of Northern Shoa during the 18th and 19th Centuries, *Rural Africana*, No. 11, 1970.

Ståhl, M., *Contradictions in Agricultural Development, A Case Study of Three Minimum Package Projects in Southern Ethiopia* (Uppsala 1973).

Taddesse Tamrat, *Church and State in Ethiopia 1270–1527* (Oxford 1972).

Tamrat Kebede, Land Tenure and Agricultural Development in Ethiopia, *Peace Corps Ethiopia Training Programme* (Addis Ababa 1970).

Taye Gulilat – Abraham, G., *Socio-economic Study of the Bako-Jere Area* (Addis Ababa 1973) (mimeo.).

Tefera Bitima, *Land Use Study of Facha* (Addis Ababa 1971) (mimeo.).

Terrefe Woldetsadik, The Unification of Ethiopia (1880–1935): Wällägä, *Journal of Ethiopian Studies*, Vol. VI, No. 1, 1968.

Trimingham, J.S., *Islam in Ethiopia* (London 1952).

Tsehai Berhaneselassie, *Menilek II, Conquest and Consolidation of the Southern Provinces* (Addis Ababa 1969) (mimeo.).

Wallensteen, P. – Schiller, B., *Fallet CADU, Organisation, Produktion, Folkligt Deltagande och Ägarstruktur i ett Regional Landsbygdsutvecklingsprojekt* (Uppsala 1972) (mimeo.).
Varghese, T.C., Land Tenure Analysis: Research Achievements and Further Areas to be Studied, *Seminar Proceedings on Agrarian Reform* (Addis Ababa 1970).
Vavilov, N.I., The Origin, Variation, Immunity and Breeding of Cultivated Plants, *Chronica Botanica,* Vol. 13, 1949/50.
Wetterhall, H., *Memo on Government Land in Ethiopia* (Addis Ababa 1968) (mimeo.).
Wickström, B., *Increasing Efficiency in the National Grain Marketing System of Ethiopia – a Policy Guideline* (EPID Publication No. 14, 1973).
Wolf, E., *Peasants* (New Jersey 1966).
Worku Moges, *Land Reform in Relation to Agricultural Development in Ethiopia* (Addis Ababa 1968) (mimeo.).
Yoseph Haile, *Relationship between Government Land and Political Power in Ethiopia* (Addis Ababa 1973) (mimeo.).

Government Documents

Ethiopian:

CADU, *Report No. 1 on the Establishment of a Regional Development Project in Ethiopia. Part II – Project Outline* (CADU Publication no. 1, 1966).
CADU, *Plan of Operation for Regional Agricultural Development in the Chilalo Awraja, Arussi Province, Ethiopia 1967–70. Annex to the Agreement of September 8, 1967 between the Swedish Government and the Imperial Ethiopian Government on co-operation in the field of regional agricultural development* (1967).
CADU, *Tentative CADU Programme 1970–1975 (CADU Publication No. 26, 1969)*
CADU, *CADU Preliminary Final Report for the Period 1967–70* (CADU Publication No. 43, 1970).
CADU, *CADU Project Description* (Addis Ababa 1971).
CADU, *CADU – A Brief Description* (Addis Ababa 1973) (mimeo.).
CADU, *General Agricultural Survey 1972* (CADU Publication No. 82, 1973).
CADU, *Annual Reports 1971/72 and 1972/73* (CADU Publication No. 77, 1973).
Central Statistical Office, *Report on a Survey of Arussi Province* (Addis Ababa 1966).
Central Statistical Office, *Report on a Survey of Beghemder Province* (Addis Ababa 1968).
Central Statistical Office, *Report on a Survey of Gemu Gofa Province* (Addis Ababa 1967).
Central Statistical Office, *Report on a Survey of Godjam Province* (Addis Ababa 1966).
Central Statistical Office, *Report on a Survey of Hararge Province* (Addis Ababa 1968).
Central Statistical Office, *Report on a Survey of Illubabor Province* (Addis Ababa 1968).
Central Statistical Office, *Report on a Survey of Kafa Province* (Addis Ababa 1968).
Central Statistical Office, *Report on a Survey of Shoa Province* (Addis Ababa 1966).
Central Statistical Office, *Report on a Survey of Sidamo Province* (Addis Ababa 1968).
Central Statistical Office, *Report on a Survey of Tigre Province* (Addis Ababa 1967).
Central Statistical Office, *Report on a Survey of Wollega Province* (Addis Ababa 1967).
Central Statistical Office, *Report on a Survey of Wollo Province* (Addis Ababa 1967).

Central Statistical Office, *Statistical Abstract 1963* (Addis Ababa 1963).
Central Statistical Office, *Statistical Abstract 1970* (Addis Ababa 1971).
Central Statistical Office, *Statistical Abstract 1971* (Addis Ababa 1972).
Central Statistical Office, *Urbanization in Ethiopia. Statistical Bulletin No. 9* (Addis Ababa 1972).
EPID, *A Master Plan for the Extension and Project Implementation Department of the Ministry of Agriculture* (EPID Publication No. 1, 1970).
EPID, *Plan of Operation for EPID 1971/72 – 1974/75* (EPID Publication No. 3, 1971).
EPID, *Manual for the Distribution of Inputs in 1973* (1972) (mimeo.).
EPID, *Work Programme and Budget for EPID 1973/74 and 1974/75* (EPID Publication No. 11, 1973).
EPID, *Annual Report for 1971/72 Financial Year* (EPID Publication No. 13, 1973).
EPID, *Annual Report for 1972/73 Financial Year* (EPID Publication No. 15, 1973).
EPID, *EPID Project Areas by Province* (1974) (mimeo.).
Ministry of Land Reform and Administration, *Policy of the Imperial Ethiopian Government on Agricultural Land Tenure* (Addis Ababa 1972) (draft).
Ministry of Land Reform and Administration, *A Policy Oriented Study of Land Settlement* Vol. I, II, (Addis Ababa 1969).
Ministry of Land Reform and Administration, *Report on Land Tenure Survey of Welega Province* (Addis Ababa 1968).
Negarit Gazeta, *Proclamation No. 117 of 1951. A Proclamation to Amend the Land Tax Proclamation of 1944.*
Negarit Gazeta, *Legal Notice No. 154 of 1951.*
Negarit Gazeta, *Proclamation No. 230 of 1966. A Proclamation to Amend the Land Tax Proclamation of 1944.*
Negarit Gazeta, *Proclamation no. 241 of 1966. A Proclamation to Provide for the Formation of Co-operative Societies.*
Negarit Gazeta, *Proclamation No. 242 of 1966. A Proclamation to Provide for the Encouragement of Capital Investment in Ethiopia.*
Negarit Gazeta, *Proclamation No. 255 of 1967. A Proclamation to Amend the Income Tax Proclamation, 1961.*
A Proclamation to Provide for the Regulation of Agricultural Tenancy Relationships. Conquering Lion of the Tribe of Judah, Haile Selassie I, Elect of God, Emperor of Ethiopia (1968) (draft).
A Proclamation to Provide for the Regulation of Agricultural Tenancy Relationships. Conquering Lion of the Tribe of Judah, Haile Selassie I, Elect of God, Emperor of Ethiopia (1970) (draft).
A Proclamation to Provide for the Regulation of Agricultural Tenancy Relationships. Haile Selassie I, Elect of God, Emperor of Ethiopia (1971) (draft).
Third Five Year Development Plan (1968–1973) (Addis Ababa 1968).

Swedish:

Bistånd Genom SIDA 1972/73 Verksamhetsberättelse (Stockholm 1974).
Kungl. Maj:ts proposition nr. 100 av år 1962. Kungl. maj:ts proposition till riksdagen angående svenskt utvecklingsbistånd, given Stockholms slott den 23 februari 1962.

Publications of the
Political Science Association in Uppsala
Edited by Axel Brusewitz

1. *Axel Brusewitz*, Studier över riksdagen och utrikespolitiken. I. Hemliga utskottet i 1809 års författning. 144 pp. 1933.
2. *Axel Brusewitz*, Studier över riksdagen och utrikespolitiken. II. Nordiska utrikesnämnder i komparativ belysning. 266 pp. 1933.
3. *Gunnar Heckscher*, Konung och statsråd i 1809 års författning. Regeringsmaktens inre gestaltning under det nya statsskickets första halvsekel. 484 pp. 1933.
4. De nordiska ländernas statsråd. Studier över deras sociala sammansättning. Utarbetade inom Seminariet för statskunskap i Uppsala. 234 pp. 1935.
5. *Sven Lindman*, Parlamentarismens införande i Finlands statsförvaltning. 85 pp. 1935.
6. *Elis Håstad*, Regeringssättet i den schweiziska demokratien. 735 pp. 1936.
7. Studier över den svenska riksdagens sociala sammansättning. Utarbetade inom Seminariet för statskunskap i Uppsala. 233 pp. 1936.
8. *Ragnar Simonsson*, Millerands presidentur. En studie över presidentmakt och parlamentarism i Frankrike. 319 pp. 1938.
9. *Gunnar Heckscher*, Svensk konservatism före representationsreformen. I. Den historiska skolans genombrott i Sverige. 257 pp. 1939.
10. *Arne Björnberg*, Parlamentarismens utveckling i Norge efter 1905. 400 pp. 1939.
11. *Arne Wåhlstrand*, 1905 års ministärkriser. 434 pp. 1941.
12. Festskrift till professor skytteanus Axel Brusewitz. Utgiven till 60-årsdagen den 9 juni 1941 av vänner och lärjungar. 506 pp. 1941.
13. *Axel Brusewitz*, Studier över riksdagen och utrikespolitiken. III. Utrikesfrågors behandling i den svenska riksdagen. 155 pp. 1941.
14. *Sixten Björkblom*, Södermanlands läns landsting. Sammansättning, organisation och verksamhet. 467 pp. ill. 1942.
15. *Edvard Thermænius*, Adressdebatten i underhuset. En studie i parlamentarisk teknik och tradition. 236 pp. 1942.
16. *Lennart Linnarsson*, Riksrådens licentiering. En studie i frihetstidens parlamentarism. 360 pp. 1943.
17. *Gunnar Heckscher*, Svensk konservatism före representationsreformen. II. Doktrin och politik 1840—1865. 271 pp. 1943.
18. *Torsten Petré*, Svenska regeringen och de norska frågorna 1884—1886. 132 pp. 1944.
19. *Olle Nyman*, Krisuppgörelsen mellan socialdemokraterna och bondeförbundet 1933. 114 pp. 1944.
20. Statsvetenskapliga studier. Till Statsvetenskapliga föreningens i Uppsala tjugofemårsdag 7 nov. 1944. Av medlemmar. 701 pp. 1944.
21. *Torsten Petré*, Ministären Themptander. 305 pp. 1 p. ill. 1945.
22. *Carl-Henrik Höjer*, Le régime parlementaire belge de 1918 à 1940. 373 pp. 1946.
23. *Ragnar Edenman*, Socialdemokratiska riksdagsgruppen 1903—1920. En studie i den svenska riksdagens partiväsen. 298 pp. 1 p. ill. 1946.
24. *Arne Wåhlstrand*, Allmänna valmansförbundets tillkomst. 104 pp. 1946.
25. *Gunnar Gerdner*, Det svenska regeringsproblemet 1917—1920. Från majoritetskoalition till minoritetsparlamentarism. 396 pp. 10 pp. ill. 1946.
26. *Nils Andrén*, Den klassiska parlamentarismens genombrott i England. 361 pp. 9 pp. ill. 1947.
27. *Olle Nyman*. Svensk parlamentarism 1932—1936. Från minoritetsparlamentarism till majoritetskoalition. 570 pp. 14 pp. ill. 1947.
28. *Arne Wåhlstrand*, Regeringsskiftena 1900 och 1920. 238 pp. 10 pp. ill. 1947.

Publications of the
Political Science Association in Uppsala
Edited by C.A. Hessler

30. *Gert Hornwall,* Regeringskris och riksdagspolitik 1840–1841. 465 pp. 6 pp. ill. 1951.
31. *Eyvind Bratt,* Småstaterna i idéhistorien. En studie i äldre statsdoktriner. 231 pp. 1951.
32. *Ragnar Andersson,* Svenska Dagbladet och det politiska livet 1897–1918. 460 pp. 3 pp. ill. 1952.
33. *Arne Wåhlstrand,* Karlstadkonferensen 1905. Protokoll och aktstycken. 474 pp. 4 pp. ill. 1953.
34. *Sten-Sture Landström,* Svenska ämbetsmäns sociala ursprung. 179 pp. 1954.
35. *Gunnar Gerdner,* Parlamentarismens kris i Sverige vid 1920-talets början. 568 pp. Ill. 1954.
36. *Carl Arvid Hessler,* Stat och religion i upplysningstidens Sverige. 242 pp. 1956.
37. *Olle Nyman,* Högern och kungamakten 1911–1914. Ur borggårdskrisens förhistoria. 408 pp. 1957.
38. *Elias Berg,* The historical thinking of Charles A. Beard. 83 pp. 1957.
39. *Olle Gellerman,* Staten och jordbruket 1867–1918. 391 pp. 1958.
40. *Olle Nyman,* Der westdeutsche Föderalismus. Studien zum Bonner Grundgesetz. 266 pp. 1960.
41. *Mats Kihlberg* och *Donald Söderlind,* Två studier i svensk konservatism 1916–1922. 268 pp. 1961.
42. *Nils Elvander,* Harald Hjärne och konservatismen. Konservativ idédebatt i Sverige 1865–1922. 530 pp. 1961.
43. *Clarence Nilsson,* Sam Stadener som kyrkopolitiker. 392 pp. 1964.
44. *Carl Arvid Hessler,* Statskyrkodebatten. 462 pp. 1964.
45. *Olof Wennås,* Striden om latinväldet. Idéer och intressen i svensk skolpolitik under 1800-talet. 412 pp. 1966.
46. *Leif Lewin,* Planhushållningsdebatten. 564 pp. 1967.
47. *Artur Mägi,* Das Staatsleben Estlands während seiner Selbständigkeit. I. Das Regierungssystem. 327 pp. 1967.
48. *Lars Rudebeck,* Party and People. A Study of Political Change in Tunisia. 275 pp. 1967.
49. *Bo Jonsson,* Staten och malmfälten. En studie i svensk malmfältspolitik omkring sekelskiftet. 388 pp. 1969.
50. Idéer och ideologier. Sex studier utgivna till Statsvetenskapliga föreningens i Uppsala femtioårsdag den 7 november 1969. 375 pp. 1969.
51. *Terry Carlbom,* Högskolelokaliseringen i Sverige 1950–1965. En studie i politiskt inflytande. 309 pp. 1970.
52. *Leif Lewin,* Folket och eliterna. En studie i modern demokratisk teori. 260 pp. 1970.
53. *Reidar Larsson,* Theories of Revolution. From Marx to the First Russian Revolution. 381 pp. 1970.
54. *Sverker Gustavsson,* Debatten om forskningen och samhället. En studie i några teoretiska inlägg under 1900-talet. 359 pp. 1971.
55. *Ragnar Simonsson,* Frankrikes författningar. 617 pp. 1971.
56. *Leontios Ierodiakonou,* The Cyprus Question. 311 pp. 1971.
57. *Evert Vedung,* Unionsdebatten 1905. En jämförelse mellan argumenteringen i Sverige och Norge. 475 pp. 1971.
58. *Hans Sjöberg,* De Gaulles Europatankar. 148 pp. 1972.
59. *Georg Andrén,* Den unge Churchill. 100 pp. 1972.
60. *Leif Lewin, Bo Jansson, Dag Sörbom,* The Swedish Electorate 1887–1968. 293 pp. 1972.
61. *Barry Holmström,* Koreakriget i svensk debatt. 334 pp. 1972.

Publications of the
Political Science Association in Uppsala
Edited by Leif Lewin

62. *Leif Lewin,* Statskunskapen, ideologierna och den politiska verkligheten. 28 pp. 1972.
63. *Jan Åke Dellenbrant,* Reformists and Traditionalists. A Study of Soviet Discussions about Economic Reform, 1960–1965. 163 pp. 1972.
64. *Peter Wallensteen,* Structure and War. On International Relations, 1920–1968. 269 pp. 1973.
65. *Georg Landberg,* Hans Järta som landshövding i Kopparbergs län 1812–1822. 295 s. 1973.
66. *Bengt Bogärde,* Svenska folkhögskolans lärarförening och staten 1902–1970. En studie i organisationsinflytande. 187 s. 1974.
67. *Michael Ståhl,* Ethiopia: Political Contradictions in Agricultural Development. 186 pp. 1974.